Governors State University

Library

Hours:

Monday thru Thursday 8:30 to 10:30

Friday and Saturday 8:30 to 5:00

Sunday 1:00 to 5:00 (Fall and Winter Trimester Only)

DEMCO

Security in
Sensor Networks

Security in Sensor Networks

Edited by Yang Xiao

Auerbach Publications
Taylor & Francis Group
Boca Raton   New York

Auerbach Publications is an imprint of the
Taylor & Francis Group, an informa business

Auerbach Publications
Taylor & Francis Group
6000 Broken Sound Parkway NW, Suite 300
Boca Raton, FL 33487-2742

© 2007 by Taylor & Francis Group, LLC
Auerbach is an imprint of Taylor & Francis Group, an Informa business

No claim to original U.S. Government works
Printed in the United States of America on acid-free paper
10 9 8 7 6 5 4 3 2 1

International Standard Book Number-10: 0-8493-7058-2 (Hardcover)
International Standard Book Number-13: 978-0-8493-7058-8 (Hardcover)

Library of Congress Cataloging-in-Publication Data

Security in sensor networks / edited by Yang Xiao.
    p. cm.
  Includes bibliographical references and index.
  ISBN 0-8493-7058-2 (alk. paper)
  1. Sensor networks--Security measures. I. Xiao, Yang.

TK7872.D48S43 2006
681'.2--dc22   2006042733

Visit the Taylor & Francis Web site at
http://www.taylorandfrancis.com

and the Auerbach Web site at
http://www.auerbach-publications.com

Dedication

To my dear father Xiao, Xiling
September 26, 1929–August 30, 2005

Contents

SECTION IV: SECURE ROUTING

SECTION V: SECURE AGGREGATION, LOCATION, AND CROSS-LAYER

About the Editor

Yang Xiao worked at Micro Linear as a MAC (Medium Access Control) architect involving the IEEE 802.11 standard enhancement work before joining the Department of Computer Science at The University of Memphis in 2002. Dr. Xiao is currently with the Department of Computer Science at the University of Alabama. Dr. Xiao is an IEEE Senior member. He was a voting member of the IEEE 802.11 Working Group from 2001 to 2004. He currently serves as Editor-in-Chief for the *International Journal of Security and Networks (IJSN)* and for the *International Journal of Sensor Networks (IJSNet)*. He serves as an associate editor or on editorial boards for the following refereed journals: *International Journal of Communication Systems, Wireless Communications and Mobile Computing (WCMC), EURASIP Journal on Wireless Communications and Networking*, and *International Journal of Wireless and Mobile Computing*. He served as a (lead) guest editor for the *International Journal of Security in Networks (IJSN)*, Special Issue on "Security Issues in Sensor Networks" in 2005; as a (lead) guest editor for *EURASIP Journal on Wireless Communications and Networking*, Special Issue on "Wireless Network Security" in 2005; as a (sole) guest editor for the *Computer Communications Journal*, Special Issue on "Energy-Efficient Scheduling and MAC for Sensor Networks, WPANs, WLANs, and WMANs" in 2005; as a (lead) guest editor for the *Journal of Wireless Communications and Mobile Computing*, Special Issue on "Mobility, Paging and Quality of Service Management for Future Wireless Networks" in 2004; as a (lead) guest editor for the *International Journal of Wireless and Mobile Computing*, Special Issue on "Medium Access Control for WLANs, WPANs, Ad Hoc Networks, and Sensor Networks" in 2004; and as an associate guest editor for the *International Journal of High Performance Computing and Networking*, Special Issue on "Parallel and Distributed Computing, Applications and Technologies" in 2003. He serves as co-editor for seven edited books: *Security in Sensor Networks, Wireless/Mobile Network Security, Adaptation Techniques in Wireless Multimedia Networks, Wireless LANs and Bluetooth, Security and Routing in Wireless Networks, Ad Hoc and Sensor Networks,* and *Design and Analysis of Wireless Networks.* He serves as a referee/reviewer for many funding agencies, as well as a panelist for NSF. His research interests include security and wireless networks. E-mail: yangxiao@ieee.org.

Contributors

Michael Chorzempa
The Bradley Department of Electrical
and Computer Engineering
Virginia Polytechnic Institute
and State University
Blacksburg, Virginia USA

Sajal K. Das
Center for Research in Wireless
Mobility and Networking
Department of Computer Science
and Engineering
The University of Texas–Arlington
Arlington, Texas USA

Tassos Dimitriou
Athens Information Technology
Athens, Greece

Mohamed Eltoweissy
The Bradley Department of Electrical
and Computer Engineering
Virginia Polytechnic Institute
and State University
Blacksburg, Virginia USA

Jun Fung
Department of Computer Science
University of Manitoba
Winnipeg, Manitoba
Canada

Y. Thomas Hou
The Bradley Department of Electrical
and Computer Engineering
Virginia Polytechnic Institute
and State University
Blacksburg, Virginia USA

Fei Hu
Computer Engineering Department
Rochester Institute of Technology
(RIT)
Rochester, New York USA

Yixin Jiang
Department of Computer Science
and Technology
Tsinghua University
Beijing, China

Farinaz Koushanfar
Electrical Engineering and Computer
Science Department
University of California–Berkeley
Berkeley, California USA

Ioannis Krontiris
Athens Information Technology
Athens, Greece

Chuang Lin
Department of Computer Science
   and Technology
Tsinghua University
Beijing, China

Yonghe Liu
Center for Research in Wireless
   Mobility and Networking
Department of Computer Science
   and Engineering
The University of Texas–Arlington
Arlington, Texas USA

Jelena Mišić
Department of Computer Science
University of Manitoba
Winnipeg, Manitoba
Canada

Vojislav B. Mišić
Department of Computer Science
University of Manitoba
Winnipeg, Manitoba
Canada

Jung-Min Park
The Bradley Department of
   Electrical and Computer
   Engineering
Virginia Polytechnic Institute
   and State University
Blacksburg, Virginia USA

Miodrag Potkonjak
Computer Science Department
University of California–Los Angeles
Los Angeles, California USA

Karthikeyan Ravichandran
Department of CSEE
University of Maryland,
Baltimore, Maryland USA

Krishna Sankar
Global Commerce Technology
   Division
CISCO Systems, Inc.
Silicon Valley, California USA

Xuemin (Sherman) Shen
Department of Electrical
   and Computer Engineering
University of Waterloo
Waterloo, Ontario
Canada

Minghui Shi
Department of Electrical
   and Computer Engineering
University of Waterloo
Waterloo, Ontario
Canada

Swapna Shroff
Department of Computer Science
The University of Memphis
Memphis, Tennessee USA

Waqaas Siddiqui
Computer Engineering Department
Rochester Institute of Technology
Rochester, New York USA

Krishna M. Sivalingam
Department of CSEE
University of Maryland
Baltimore, Maryland USA

Xu (Kevin) Su
Department of Computer Science
University of Texas–San Antonio
San Antonio, Texas USA

Yu-Chee Tseng
Department of Computer Science
National Chiao-Tung University
Hsin-Chu, Taiwan

Xudong Wang
Kiyon, Inc.
La Jolla, California USA

You-Chiun Wang
Department of Computer Science
National Chiao-Tung University
Hsin-Chu, Taiwan

Ming Chuen (Derek) Wong
Department of Computer Science
The University of Memphis
Memphis, Tennessee USA

Mingbo Xiao
Xiamen University
Xiamen, Fujian Province
China

Yang Xiao
Department of Computer Science
The University of Alabama
Tuscaloosa, Alabama USA

Guangsong Yang
Xiamen University
Xiamen, Fujian Province
China

Wei Zhang
Center for Research in Wireless
  Mobility and Networking
Department of Computer Science
  and Engineering
The University of Texas–Arlington
Arlington, Texas USA

Wensheng Zhang
Department of Computer Science
Iowa State University
Ames, Iowa USA

Sencun Zhu
Department of Computer Science
  and Engineering
The Pennsylvania State University
University Park, Pennsylvania USA

Preface

Sensor networks have many applications, and security issues in some sensor applications are very important, such as monitoring applications in the battlefield. Sensor networks differ from other traditional networks in many aspects, such as limited energy, limited memory space, limited computation capability, etc. Therefore, sensor network security has some unique features that do not exist in other networks. For example, a sensor node has very limited memory space so that the number of keys that can be stored in the memory is very limited.

The goal of the book is to serve as a useful reference for researchers, educators, graduate students, and practitioners in the field of security in sensor networks. In meeting its goal, this book covers all areas of sensor security.

This book contains 13 invited chapters from prominent sensor security researchers working around the world. The book is divided into five sections: I: Attacks; II: Encryption, Authentication, and Watermarking; III: Key Management; IV: Secure Routing; and V: Secure Aggregation, Location, and Cross-Layer.

Of course, the represented topics are not an exhaustive representation of the world of security in sensor networks. Nonetheless, they represent the rich and useful strategies and contents that we have the pleasure of sharing with the readers.

This book has been made possible due to the efforts and contributions of many people. First and foremost, we would like to thank all the contributors for their tremendous effort in putting together excellent chapters that are comprehensive and very informative. We would also like to thank Rich O'Hanley as well as other staff members at Taylor & Francis for putting this book together. Finally, I would like to dedicate this book to my father, who passed away during the period of my editing this book.

Yang Xiao

ATTACKS

I

Chapter 1

Attacks and Defenses of Routing Mechanisms in Ad Hoc and Sensor Networks

You-Chiun Wang and Yu-Chee Tseng

Contents

1.1 Abstract

An ad hoc or a sensor network is composed of a collection of wireless nodes. A node can only communicate with other nodes within its limited transmission range. To facilitate communication between two nodes without a direct communication link, routing protocols must be developed to support multi-hop communication. Although many routing protocols have been proposed for ad hoc and sensor networks, most of them assume that other nodes are trustable and thus do not consider the security and attack issues. To assure a source node of finding a route to its destination, most routing protocols try to invite all available nodes to participate in the routing mechanism. This provides myriad opportunities for attackers to destroy the routing mechanism. In this chapter, we briefly introduce some existing routing protocols, discuss the weaknesses of these protocols and possible types of attacks, and provide a comprehensive survey of recent research on defense approaches to these attacks.

1.2 Introduction

Recently, the rapid development of wireless communication not only alleviates the wired problem of traditional networks, but also provides the capability of mobile communication and ubiquitous computing. The ad hoc network architecture is a representative example and has been proposed to rapidly set up a network when needed [10]. An ad hoc network consists of a collection of wireless nodes. Each node can directly communicate with other nodes within its transmission range. Communication between out-of-range nodes has to be routed through one or multiple intermediate nodes. Thus, each node also acts as a router. Because nodes can be mobile, the corresponding protocol should be able to handle rapid topology change.

A sensor network is also considered an ad hoc network in which nodes are extended with sensing capability. Such a network is composed of one or multiple *remote sinks* and many tiny, low-power *sensor nodes*, each containing some actuators, sensing devices, and a wireless transceiver [1]. These sensor nodes are massively deployed in a region of interest to gather environmental information, which is to be reported to remote sinks. It thus provides an inexpensive and powerful means to monitor the physical environment. The functionalities of a remote sink are to collect data from sensor nodes and to transmit queries or commands to sensor nodes. However, a sensor network differs from an ad hoc network in several aspects. First, a node in an ad hoc network is usually a laptop or a PDA, while a sensor node is typically a smaller device with a low-speed processor, limited memory, and a short-range transceiver. Thus, protocols/algorithms running on a sensor node should be simple. Second, because a sensor node is typically powered by batteries, energy consumption is a more critical design issue in a sensor network. Third, the communication patterns in sensor networks may differ from those in ad hoc networks. Fourth, sensor nodes are relatively less mobile than those in ad hoc networks.

Because the transmission between two out-of-range nodes has to rely on relay nodes, many routing protocols have been proposed for ad hoc and sensor networks. However, most of them assume that other nodes are trustable and thus do not consider the security and attack issues. To assure a source node of finding a routing path to its destination, most routing protocols [19,32,33] attempt to invite all available nodes in the network to participate in the routing mechanism. This provides many opportunities for attackers to break the network.

Although many security solutions have been proposed for wire-line networks, they may not be directly applied to ad hoc and sensor networks. The difficulties and challenges are listed as follows [10]:

1. Because the transmission medium is open, ad hoc and sensor networks are more vulnerable to physical security threats than wireline networks. Possible physical attacks range from passive eavesdropping to active interference. In addition, nodes in the network without adequate protection may be captured, compromised, and hijacked by the adversary. In this case, the authentication information is disclosed and the adversary can use these hijacked nodes to disrupt the network.

2. Authentication relies on public key cryptography or certification authorities may be difficult to accomplish in an ad hoc network because such a network does not have any centralized network infrastructure. In addition, cryptography that needs complicated computation or large memory space cannot be performed on sensor nodes.

3. Due to the lack of centralized network infrastructure, a node in an ad hoc network has to detect the possible attacks by itself or cooperates with its neighbors to find out potential attackers. The effect is usually limited because nodes can only obtain local information. Furthermore, an attacker may claim other legitimate nodes to be illegal. For sensor networks, some centralized intrusion detection schemes may be applied in the remote sink. However, this will drastically increase traffic between sensor nodes and the remote sink.

4. Any security solution with a static configuration may not be suitable for ad hoc networks because nodes have mobility and the network topology may change frequently. Nodes have to continuously detect possible attackers because their neighbors are not fixed. Similarly, for a sensor network, a malicious node with mobility can roam in the network and attack different parts of the network.

5. Because nodes normally rely on batteries to provide energy, an attacker can frequently flood fake or dummy messages to exhaust other nodes' energies. An attacker can disguise its packets as normal ones or replay other nodes' packets to waste energy of normal nodes.

In this chapter, we present a survey of possible attacks and existing defense schemes of routing mechanisms in ad hoc and sensor networks. Section 1.3 briefly introduces some routing protocols in ad hoc and sensor networks. Section 1.4 discusses several possible attacks to these routing protocols. A survey of defense schemes is presented in Section 1.5. Conclusions are drawn in Section 1.6.

1.3 Routing Protocols in Ad Hoc and Sensor Networks

To support multi-hop communication in ad hoc networks, many routing protocols have been designed [12]. These protocols can be classified into *table-driven* (or *proactive*) and *on-demand* (or *reactive*) ones. In table-driven routing protocols, nodes need to exchange routing information regardless of communication requests. Such protocols attempt to maintain consistent, up-to-date routing information for each node to reach every other node in the network. Therefore, they require each node to maintain one or more tables to store routing information. Additionally, any change in network topology may need to be propagated to the whole network to maintain a consistent network view. The *destination-sequenced distance-vector (DSDV)* routing protocol [32] is a representative example. In DSDV, each node maintains a *forwarding table* in which each entry contains a destination address, the next hop to the destination, the number of hops to the destination, and a sequence number. Nodes will periodically exchange the contents of their forwarding tables. To relay packets, an intermediate node simply has to look its forwarding table to find out the next hop to the destination. Other table-driven protocols include the *topology broadcast based on reversed path forwarding (TBRPF)* routing protocol [4], the *optimized link state routing (OLSR)* protocol [18], and the *fisheye state routing (FSR)* protocol [31].

On-demand protocols are more popular for ad hoc routing [12]. The main feature of such protocols is that nodes exchange routing information only when there are communications awaiting. This can reduce routing overhead compared to the table-driven protocols. When a node attempts to communicate with another node, it floods a *route request (RREQ)* packet in the network. Nodes receiving the RREQ packet will send back a *route reply (RREP)* packet to the source if they know how to route to the destination; otherwise, they forward the RREQ packet to other nodes. There are several variations of on-demand routing. The *dynamic source routing (DSR)* protocol [19] is derived based on *source routing*. On receiving an RREQ, an intermediate node inserts its address into the packet and rebroadcasts it. Therefore, the destination will have the entire path from the source to itself. The destination then responds to an RREP with the entire routing path to the source. The whole routing path is indicated in each data packet initiated by the source. The *ad-hoc on-demand distance vector routing (AODV)* protocol [33] also finds routes by flooding RREQ packets. However, unlike DSR, each intermediate node maintains a forwarding table to indicate the next node leading to the destination so that the source does not need to insert the whole routing path in data packets. Other on-demand protocols include the *lightweight mobile routing (LMR)* protocol [8], the *temporally order routing algorithms (TORA)* [30], and the *associativity-based routing (ABR)* protocol [36].

Several protocols assume that each node is equipped with a *global positioning system (GPS)* device to provide the node's geographic location. By knowing the approximate location of the destination, the source can limit the flooding area of its RREQ packet. For example, the *location-aided routing (LAR)* protocol [22] creates a *request zone*, which contains the expected zone of the destination, as shown in Figure 1.1. Only intermediate nodes inside the request zone will forward the RREQ packet, so the overhead of flooding can be reduced. Other examples using geographic information include the *distance routing effect algorithm for mobility (DREAM)* [3], the *greedy perimeter stateless routing (GPSR)* protocol [21], and the *geographic addressing and routing (GeoCast)* protocol [28].

Routing protocols designed for ad hoc networks typically support routing between any pair of nodes. However, sensor networks have more specialized communication patterns. There are three common categories [20]:

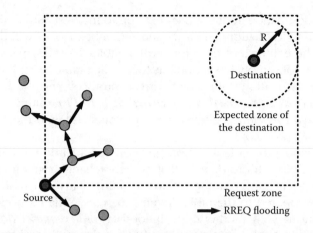

Figure 1.1   Limited flooding of RREQ packets in LAR. *R* is the maximum distance that the destination can move from its previous known location.

1. **Many to one:** Multiple or all sensor nodes report their collected data to a remote sink.
2. **One to many:** The remote sink multicasts or broadcasts a query or a command to multiple or all sensor nodes.
3. **Local communication:** Neighboring sensor nodes send localized messages to each other.

The traditional ad hoc routing protocols may not be suitable for sensor networks. In particular, a sensor network usually forms a simple spanning tree rooted at the remote sink for the routing purpose.

1.4   Attacks on Routing Mechanisms

Most routing protocols designed for ad hoc and sensor networks assume that nodes in the network do not misbehave. This provides many opportunities for malicious nodes to attack and disrupt the routing mechanism. These attacks can be classified as *passive* or *active* [10,27]. A passive attack typically involves only eavesdropping the routing traffic to discover valuable information. It is usually difficult to detect passive attacks because such attacks do not destroy the operations of routing protocols. Two possible solutions can be used to restrain eavesdropping. One is to adopt encryption or other security mechanisms in the application layer [20]. The other solution is to transmit parts of a message over multiple disjoint paths and reassemble them at the destination [5].

An active attack may attempt to disrupt the routing mechanism, intentionally modify the routing messages, gain authentication or authorization, or even control the whole network by generating false packets into the network or by modifying or dropping legitimate packets sent by other nodes. Active attacks can be further categorized into *external* and *internal* attacks. An external attack is generated by malicious nodes that do not belong to the network. An internal attack is caused by compromised or hijacked nodes that were formerly legitimate. Security mechanisms that rely on authentication or encryption may not handle internal attacks because these compromised nodes also have the keys and thus are treated as authorized parties in the network.

For sensor networks, Karlof and Wagner [20] classify the attackers into *sensor-class* and *laptop-class* attackers according to their capabilities. A sensor-class attacker has the similar capability of other legitimate nodes in the network, so the attacks caused by these attackers are less complicated. A laptop-class attacker has more powerful computing devices, more battery power, larger memory, and even high-power radio transmitters. Attacks caused by laptop-class attackers can be different from those in ad hoc networks. For example, a laptop-class attacker may be able to jam a large range of or even the entire sensor network, while an attacker in an ad hoc network is only able to jam the radio links in its vicinity.

Below, we introduce several possible active attacks in ad hoc and sensor networks. We divide these attacks into three classes: (1) attacks on route discovery process, (2) attacks on route selection process, and (3) attacks after establishing routing paths.

1.4.1 Attacks on Route Discovery Process

Such attacks attempt to prevent other legitimate nodes from establishing routing paths by sending fake routing information. Moreover, a malicious node can send excessive route request messages to exhaust the network bandwidth. The former is to provide fake information to spoof the route discovery process, while the latter is to overly use the route discovery process. Both of them attempt to cause *denial of service (DoS)*.

1.4.1.1 Fake Routing Information

A straightforward attack against a routing protocol is to provide fake routing information during its route discovery phase. For table-driven routing protocols, a malicious node can interfere with other legitimate nodes by advertising incorrect routing information to invalidate their routing tables (or forwarding tables) [27]. For on-demand routing protocols, a malicious node can reply a nonexisting route to the source or alter the addresses in an RREQ packet to spoof the destination. It can also modify an RREP packet to cause invalid route to the source. A malicious node can also silently drop all RREQ and RREP packets passing through it to refuse participating in the route discovery process.

1.4.1.2 Rushing Attacks

A *rushing attack* [16] mainly intends to break the route discovery process of an on-demand routing protocol. In an on-demand routing protocol, a node that wants to establish a route to a destination floods the network with RREQ packets. To reduce the overhead of flooding, each node typically forwards only the first-arrived RREQ packet originating from the source. Such property is exploited by a rushing attacker. In particular, after an attacker receives the RREQ packet from the source, it propagates its modified RREQ packet to other intermediate nodes *before* legitimate RREQ packets reach them. When the legitimate RREQ packets are received, they may be discarded. In this case, the attacker can prevent a valid path from being established or increase its chance of being selected as part of a routing path.

Note that a rushing attack can succeed only if the attacker can send its modified RREQ packets to other nodes before these nodes receive the legitimate ones. To achieve this, the attacker can reduce delays at either the MAC or routing layer to send out its packets as fast as possible. For example, the attacker can select a smaller number to run the backoff mechanism at the MAC layer after collision. A more powerful rushing attacker can utilize a wormhole [15] (refer to Section 1.4.2.3) to rush packets.

1.4.1.3 RREQ Flood Attacks

The fundamental mechanism of the route discovery process of an on-demand routing protocol is to flood RREQ packets in the network. An *RREQ flood attacker* abuses this mechanism to result in a DoS attack. In particular, the attacker can generate frequent unnecessary or false RREQ packets to make the network resources unavailable to other legitimate nodes. In addition, the attacker can make other nodes' routing tables overflow by flooding excessive RREQ packets with different or even nonexistent destinations [10], so that creation of new routes by other legitimate nodes will be prohibited.

1.4.2 Attacks on Route Selection Process

This type of attack attempts to increase the chance that malicious nodes are selected by other legitimate nodes as part of their routes. After establishing a route through itself, the attacker can overhear the transmitted messages

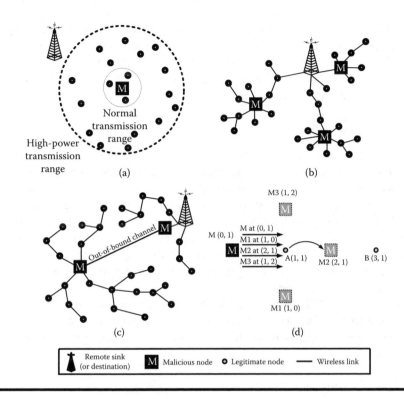

Figure 1.2   Attacks on route selection processes: (a) HELLO flood attack, (b) Sinkhole attack, (c) Wormhole attack, and (d) Sybil attack.

or combine the attacks discussed in Section 1.4.3 to disrupt the network. Figure 1.2 shows four possible attacks of this type, including HELLO flood, sinkhole, wormhole, and Sybil attacks.

1.4.2.1   HELLO Flood Attacks

Many protocols require nodes to broadcast localized HELLO messages to announce themselves to their neighbors. A node receiving such a message will assume that the sender is its one-hop neighbor. However, this assumption may be violated when there are laptop-class attackers in a sensor network. A laptop-class attacker can utilize a large transmission power to broadcast its HELLO message to cover a large range of sensors. The receiving nodes will be convinced that the attacker is their one-hop neighbor, as shown in Figure 1.2a. Such an attack is called a *HELLO flood attack* [20].

Protocols that rely on localized information exchange between neighboring nodes for topology maintenance or flow control are vulnerable to the HELLO flood attack. In addition, the attacker can advertise a higher quality or shorter route to the destination. This may even cause other nodes to follow the same route to the destination. However, most messages from legitimate nodes may not be sent to the attacker because these nodes have smaller transmission ranges.

1.4.2.2 Sinkhole Attacks

The objective of a *sinkhole attack* [20] is to attract all neighboring nodes of the attacker to establish routes through the attacker, as shown in Figure 1.2b. In this scenario, all traffic from a particular area will flow through the attacker, thus creating a metaphorical sinkhole with the attacker at the center. Sinkhole attacks typically work by making a malicious node look especially attractive to surrounding nodes with respect to the routing algorithm. For example, an attacker can advertise an extreme high quality route to some destinations, or even spoof these surrounding nodes that the attacker itself is neighboring to the destination.

Unlike the HELLO flood attack, the sinkhole attacker usually utilizes a normal transmission power. Thus, a sinkhole attacker may affect only part of the network, and both ad hoc and sensor networks are vulnerable to such attacks.

1.4.2.3 Wormhole Attacks

Multiple malicious nodes can cooperate to generate attacks against the network. One of the representative examples is the *wormhole attack* [15]. In a wormhole attack, two distant malicious nodes utilize an out-of-bound channel available only to the attackers to tunnel messages received by one side to another side. Specifically, packets transmitted through the wormhole tunnel usually have lower latency than those packets sent between the same pair of nodes over normal multi-hop routing. This will result in a false appearance that routing through these two malicious nodes is a better choice. Therefore, neighboring nodes will select the malicious nodes as the intermediate nodes in their routes. In addition, wormholes can create a fake network topology by relaying packets between two distant nodes. In this case, these two distant nodes may be considered as neighbors of each other.

A sensor network is more vulnerable to the wormhole attack due to the following two reasons. First, laptop-class attackers can utilize out-of-bound channels to create low-latency, high-bandwidth tunnels more easily because they have more powerful transceivers compared to other sensor nodes. Second, an adversary situated close to the remote sink can control a lot of routing by creating a well-placed wormhole. Figure 1.2c gives an example, where more than half of the sensor nodes will be guided to the wormhole tunnel.

1.4.2.4   Sybil Attacks

In a *Sybil attack* [29], a malicious node can disguise itself as multiple different nodes by advertising multiple identities to its neighbors. Because the Sybil attacker can create many fake nodes, it thus can increase the probability that the malicious node is selected by other nodes as part of their routing paths. Also, the Sybil attack can significantly reduce the effectiveness of fault-tolerance schemes such as multi-path routing [7,17] because other nodes will treat the fake nodes generated by the malicious node as different nodes and establish "different" routes through the malicious node.

A Sybil attacker can also spoof nodes using a geographic routing protocol, such as the GRID routing protocol [24]. Figure 1.2d gives an example. A malicious node M at actual location $(0,1)$ advertises not only its true identity and location, but also three forged nodes $M1$, $M2$, and $M3$ at locations $(1,0)$, $(2,1)$, and $(1,2)$, respectively. After receiving these advertisements, if a node A located at $(1,1)$ wants to transmit packets to another node B located at $(3,1)$, it will select the fake node $M2$ (which locates at $(2,1)$) as its forwarding node. Because the malicious node M is a neighbor of node A, it can overhear the transmission and takes a further action.

1.4.3   Attacks after Establishing Routing Paths

Once a source node establishes a route through a malicious node, the malicious node can unscrupulously drop the data packets from the source or modify the contents of packets if encryption is not applied. A malicious node can use the attacks discussed in Section 1.4.2 to insert itself to the routing path. In addition, an attacker can play as a source node to establish routes to other nodes, and then send dummy messages to exhaust their energies and the network bandwidth.

1.4.3.1   Blackhole Attacks

Multi-hop communications must rely on the cooperation of participating nodes to forward the received messages. In a *blackhole attack* [10], malicious nodes violate such an assumption by dropping all received messages from the source to prevent these messages from being propagated any further.

However, most routing protocols have the *route maintenance* mechanism such that if a node in the path finds that its next-hop neighbor no longer propagates its data packets, it will notify the source to recreate another routing path. In this case, a blackhole attack is trivially defended. A more cunning form of this attack is that the attacker selectively forwards packets [20] to cheat the source that this route is still alive.

1.4.3.2 Spam Attacks

Like spam mail, a spam attacker [35] frequently generates a large number of unsolicited and useless messages to the network. These messages will waste the network bandwidth and the energies of nodes that receive or forward these messages. The spam attacks are more jeopardous to sensor networks. This is because in a sensor network, environmental data collected by sensors will all be transmitted to the remote sink. A spam attacker can thus generate a lot of dummy messages to the sink to consume energies of relayed sensors, especially those close to the sink. Once the energies of these sensors are exhausted, the sink will never receive data from the sensor network. In this case, the whole sensor network is destroyed, even though most of sensors are alive.

1.4.4 Summary of Attacks

Here we compare attacks to ad hoc and sensor networks in Table 1.1. Recall that nodes in an ad hoc network have similar capability, while there can be powerful laptop-class attackers in a sensor network. From Table 1.1, we can observe that the fake routing information, sinkhole attacks, Sybil attacks, and blackhole attacks affect both networks. Because a sensor network usually forms a spanning tree rooted at the remote sink for the routing purpose, attacks against on-demand routing protocols, such as rushing attacks and RREQ flood attacks, may not affect sensor networks. The HELLO flood attacks can only affect sensor networks because in an ad hoc network, a malicious node may not be able to generate a large power to cover most nodes in the network. The wormhole attack is based on the assumption that two malicious nodes can utilize an out-of-bound channel to communicate with each other. In addition, they have to provide a low-latency, high-bandwidth, and long-distance link to tunnel packets

Table 1.1   Comparison of Attacks

Attack	Ad Hoc Network	Sensor Network
Fake routing information	√	√
Rushing attacks	√	
RREQ flood attacks	√	
HELLO flood attacks		√
Sinkhole attacks	√	√
Wormhole attacks	Partially	√
Sybil attacks	√	√
Blackhole attacks	√	√
Spam attacks	Partially	√

Note: A √ means that the corresponding network is vulnerable to such an attack.

Table 1.2   Comparison of Attacks on Table-Driven and On-Demand Routing Strategies

Attack	Table-Driven Protocol	On-Demand Protocol
Fake routing information	√	√
Rushing attacks		√
RREQ flood attacks		√
HELLO flood attacks	√	√
Sinkhole attacks	√	√
Wormhole attacks	√	√
Sybil attacks	√	√

Note: A √ means that the corresponding network is vulnerable to such an attack.

between them. These assumptions are sometimes difficult to accomplish in an ad hoc network. Finally, the spam attack can partially affect an ad hoc network because all nodes in an ad hoc network can become the possible destination nodes. Such an attack can succeed in a sensor network because the remote sink is usually the only destination.

Table 1.2 compares the attacks on table-driven and on-demand routing strategies. Most attacks will affect both routing strategies, except rushing attacks and RREQ flood attacks, which are only against the on-demand routing protocols.

1.5   Defense Schemes

To avoid various attacks on ad hoc and sensor networks, many defense schemes have been designed. In this section, we give a survey of these defense schemes against the attacks discussed in Section 1.4.

1.5.1   Defenses against Fake Routing Information and RREQ Flood Attacks

To prevent an external attacker from generating fake routing information or RREQ flood attacks against the network, one possible solution is to apply security mechanisms such as authentication to the routing protocol [13,14,25]. Nodes in the network share keys to authenticate their data packets and routing control messages such as RREQ and RREP. Because an external attacker does not have the keys to authenticate its packets, all its fake routing information and dummy RREQ packets will not be accepted by other legitimate nodes, so the attacks can be defended.

The *Localized Encryption and Authentication Protocol (LEAP)* [37] proposes a key management protocol for sensor networks, where different types of key managements are utilized for different security requirements. In LEAP, four types of keys are established for sensor nodes:

1. **Individual key:** Each sensor node shares a unique key with the remote sink. This key is used for secure communication between the sensor node and the remote sink.

2. **Group key:** The group key is globally shared among all sensor nodes and the remote sink. This key is used by the remote sink to encrypt broadcast messages.

3. **Cluster key:** A cluster key is shared by a sensor node and *all* its neighbors. It is mainly used for securing locally broadcast messages, such as routing control messages.

4. **Pairwise shared key:** Every sensor node shares a pairwise key with *each* of its neighbors for secure communication.

LEAP can prevent an *internal* attacker from disrupting the entire network, because a sensor node does not have a network-wide authentication key. (Note that the group key is only used to encrypt messages from the remote sink and it cannot be used for authentication.) A hijacked node can only have local keys shared with its neighbors so that it can only affect its neighbors.

1.5.2 Defenses against Rushing Attacks

A rushing attack is caused by a malicious node rapidly transmitting fake RREQ packets to invalidate the legitimate ones. To defend against such attacks, Hu et al. [16] propose the *randomized RREQ forwarding* and the *secure neighbor detection* schemes. In the randomized RREQ forwarding scheme, each node collects a number of RREQ packets and then randomly selects one of them to forward. In this case, a malicious node can take no advantage by transmitting RREQ packets very quickly, because its neighbors will wait for other legitimate RREQ packets. However, the fake RREQ packet can still be selected, so the secure neighbor detection scheme is invoked to determine whether the sender of the RREQ packet is a legitimate node. The secure neighbor detection scheme utilizes authenticated messages exchanged between two nodes to verify their identities and legitimacy.

1.5.3 Defenses against HELLO Flood Attacks

Because a HELLO flood attack is caused by a malicious node utilizing a large transmission power to generate asymmetric links between it and other legitimate nodes, one intuitive defense against such an attack is to verify the *bidirectionality* of a link between two "neighboring" nodes. The LEAP discussed in Section 1.5.1 takes this strategy. In LEAP, when a node *u*

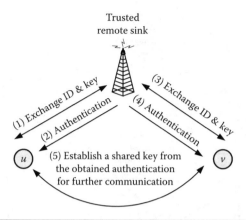

Figure 1.3 Trusted third-party authentication in a sensor network.

attempts to discover its neighbors, it broadcasts a HELLO message and waits for each neighbor v to respond with its identity. The response from v is authenticated by a *message authentication code* so that u can verify the response message. Note that node u will consider node v as its neighbor only if it receives a correct response from v. In this case, HELLO flood attacks will fail because in LEAP a node will only consider another node as its neighbor if there is two-way communication between them.

Another solution for sensor networks is to utilize the remote sink as a trusted third party to help two sensor nodes verify each other [20]. Specifically, each sensor node in the network shares a unique symmetric key with the remote sink. Two nodes u and v can then verify each other's identity and establish a shared key through the remote sink's authentication, as shown in Figure 1.3. A pair of neighboring nodes can thus use the resulting key to communicate with each other. To avoid a mobile attacker roaming around a stationary network or using the HELLO flood attack to establish shared keys with too many sensor nodes, the remote sink can reasonably limit the number of verified neighbors for each sensor node and reject a request when a node exceeds the limitation.

1.5.4 Defenses against Sinkhole Attacks

Recall that the intention of a sinkhole attacker is to attract all its neighbors to establish routes through it. To achieve this, Culpepper and Tseng [9] provide a possible trick for the attacker by generating bogus RREQ packets in the DSR protocol. A malicious node can broadcast bogus RREQ packets with properly selected sources and destinations, very high sequence numbers, and route records that specify one-hop routes from the sources to the

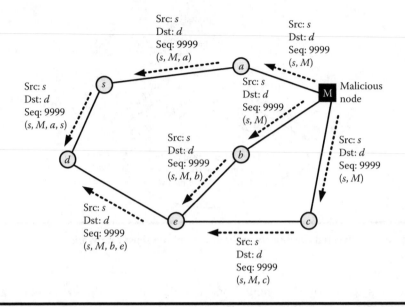

Src: *s*
Dst: *d*
Seq: 9999
(*s, M, a*)

Src: *s*
Dst: *d*
Seq: 9999
(*s, M*)

Src: *s*
Dst: *d*
Seq: 9999
(*s, M, a, s*)

Src: *s*
Dst: *d*
Seq: 9999
(*s, M*)

Malicious
node

Src: *s*
Dst: *d*
Seq: 9999
(*s, M, b*)

Src: *s*
Dst: *d*
Seq: 9999
(*s, M*)

Src: *s*
Dst: *d*
Seq: 9999
(*s, M, b, e*)

Src: *s*
Dst: *d*
Seq: 9999
(*s, M, c*)

Figure 1.4   An example of sinkhole attacks in DSR by generating bogus RREQ packets.

malicious node itself. Such RREQ packets will make a false appearance to other nodes that the malicious node is an immediate neighbor to the source, and the information is the freshest because the sequence numbers are quite large. So these nodes receiving the bogus RREQ packets will update the fake routing information into their route caches. Figure 1.4 gives an example, where the malicious node *M* broadcasts a bogus RREQ packet with source as *s* and destination as *d*. After receiving the bogus RREQ packet, nodes *a*, *b*, and *c* attempt to learn routes from the route record. By reversing the path recorded in the RREQ packet, they will falsely conclude that node *M* has a one-hop route to node *s*, and replace the old route by this fake route in their caches. Note that even if node *a* is a neighbor of node *s*, it will add this route to its cache because the sequence number in the bogus RREQ packet is larger than the sequence number for any route that node *a* previously learned to node *s*. The malicious node *M* can also fabricate RREQ packets with different combinations of sources and destinations. By repeating this attack periodically, all neighbors will believe that node *M* has the shortest path to every other node.

To detect such attacks, the *sinkhole intrusion detection system (SIDS)* [9] utilizes three indicators to determine whether there are sinkhole attackers in the network:

1. **Discontinuity of sequence numbers:** In theory, the sequence numbers of packets that originate from a node should be strictly increasing in DSR. However, a sinkhole attacker attempts to use a very large sequence number to update the contents of other nodes' route caches. Therefore, a node can monitor the sequence numbers of receiving RREQ packets and pay attention to those that are not strictly increasing (or unusually large).
2. **Ratio of verified RREQ packets:** When a node initiates an RREQ packet, the source address should be its own address. Such RREQ packets can be verified by the source's neighbors. However, a sinkhole attacker initiates RREQ packets with different sources and periodically broadcasts these bogus RREQ packets to the network. Therefore, a lower ratio of verified RREQ packets in the overall network may indicate the presence of a sinkhole attacker.
3. **Ratio of routes through a particular node:** Because a sinkhole attack causes nodes in the network to add routes through the attacker, nodes can determine the existence of a sinkhole attacker by checking their routing caches. If a node finds that most routes in its cache are through a particular node, it will suspect that this node is a potential attacker.

1.5.5 Defenses against Wormhole Attacks

Two distant malicious nodes can use an out-of-bound channel to result in a wormhole attack by tunneling packets through the channel. The concept of *packet leash* [15] is introduced to defend against such attacks. A *leash* is the information added in a packet to restrict the packet's maximum allowed transmission distance. Hu et al. [15] propose two leashes: *geographical leashes* and *temporal leashes*. The geographical leash guarantees that the receiver of the packet is within a certain distance of the sender, while the temporal leash ensures that the packet has an upper bound on its lifetime, which restricts its maximum traveling distance.

To use temporal leashes, all nodes in the network must have tightly timed synchronization. Specifically, the maximum difference between any two nodes' clocks is bounded by Δ, and this value should be known by all nodes. When a node s transmits a packet, it includes its current time t_s and utilizes authentication to protect this packet. When the node r receives this packet at time t_r, it can determine whether there are wormhole attackers in its route, based on the claimed transmission time t_s and the propagation speed v_c. In particular, node s can embed an expiration time t_{expire} in its packet; $t_{expire} = t_s + L/v_c - \Delta$, where L is the maximum distance that the packet is allowed to travel. When node r receives the packet, it will check if $t_r < t_{expire}$. If so, node r will accept the packet. Otherwise, the packet is discarded and this indicates that the earlier RREQ has been tunneled.

1.5.6 Defenses against Sybil Attacks

In a Sybil attack, a malicious node disguises as different nodes by impersonating other nodes or claiming fake identities to its neighbors. Newsome et al. [29] propose two schemes, *radio resource testing* and *random key predistribution*, are designed to defend against Sybil attacks. The radio resource testing scheme is based on the assumption that each physical node (including the attacker) has only one radio and cannot simultaneously send or receive on more than one channel. A node that attempts to check whether there are fake nodes pretended by a Sybil attacker in its neighborhood can assign each of its neighbors a different channel to broadcast messages. The node then randomly selects a channel to listen. If a message can be received, the neighbor is indicated as a legitimate node. Otherwise, the neighbor is treated as a fake node pretended by the Sybil attacker.

The random key predistribution scheme is derived from the *key pool* scheme [6, 11]. The key pool scheme randomly assigns k keys to each node from a pool of m keys. If two nodes share q common keys, they can establish a secure link. However, the key pool scheme cannot defend against the Sybil attack because if an attacker compromises multiple nodes, it can use all possible combinations of the compromised keys to generate new identities. The random key predistribution scheme solves this problem using a pseudo-random hash function to assign keys and validate the identity of a node. Specifically, let $\Omega(ID) = \{K_{\beta_1}, K_{\beta_2}, \ldots, K_{\beta_k}\}$ be the set of keys assigned to the node whose identity is ID, where $\beta_i = PRF_{H(ID)}(i)$ is the index of its ith key in the key pool, H is a one-way hash function, and *PRF* is a pseudo-random function. Because the indices of keys assigned to a node are determined by the hash value of its identity, and it is very hard to inverse the hash function to obtain the original identity, a Sybil attacker cannot just collect a set of keys and claim fake identities from these keys.

1.5.7 Defenses against Blackhole Attacks

Blackhole attacks occur when a malicious node intentionally drops all routing messages after establishing the routing path. An attacker can even selectively forward and drop packets to avoid triggering the route maintenance mechanism at a source node to select another route. To defend against such attacks, Marti et al. [26] propose a *watchdog* scheme to identify potential malicious nodes and a *path-rater* scheme to help routing protocols avoid these nodes. Using the DSR protocol, the watchdog works on the assumption that a node can overhear the packets transmitted by its neighbors. The idea of watchdog is that when a node A transmits a data packet to its next-hop neighbor *B*, node *A* will overhear the transmission from *B* to check whether node *B* has really transmitted the data packet to *B*'s next-hop neighbor. The watchdog at each node maintains a counter to record the misbehavior of each of its next-hop neighbors. Once the value of the counter exceeds a threshold, the watchdog will infer that its next-hop neighbor may be a malicious node and reports to the source. The path-rater scheme, combined with the watchdog, helps a source node select the most reliable route. Each node assigns a non-negative rating to every normal node that it knows in the network (including itself), and a highly negative rating to each malicious node. The overall rating of a routing path is the average rating of nodes on that route. The source node then selects the routing path with the highest rating to forward its packets. Note that since a malicious node is assigned a highly negative rating, a routing path with a negative rating indicates the presence of malicious nodes. Therefore, the path-rater scheme can help a node select a route without malicious nodes.

Another solution for blackhole attacks is proposed [2] that utilizes the following four mechanisms:

1. **Source routing:** The source specifies in each data packet the sequence of nodes that the packet has to traverse.
2. **Destination acknowledgments:** The destination sends back an *acknowledgment (ACK)* to the source along the same route (but in reverse direction) when it receives each data packet.
3. **Timeouts:** The source and each intermediate node set a timer for each data packet, during which they expect to receive an ACK from the destination or a *fault announcement (FA)* from other intermediate nodes.
4. **Fault announcements:** When the timer expires at a node, it generates an FA and propagates the FA back to the source.

All data, ACK, and FA messages are authenticated by a message authentication code so that these messages cannot be modified or fabricated by a malicious node. Note that the source can detect the presence of a potential blackhole attacker when it receives an FA message and thus select another route to forward its packets.

1.5.8 Defenses against Spam Attacks

Spam attacks are caused by malicious nodes generating frequent dummy messages to specified targets in the network. A sensor network is more vulnerable to such attacks because the remote sink is usually the only target to be attacked. To defend against spam attacks, the *detect and defend spam (DADS)* scheme [35] proposes a concept of *quarantine regions* to "isolate" spam attackers. In DADS, the remote sink is responsible for detecting whether there are spam attacks in the network. The remote sink can detect spam attacks by the three methods. The first one is to filter incoming messages according to their contents and detect nodes that send faulty message frequently. The second method uses the frequencies of messages sent by the sensor nodes in the same region. The third method is to observe the packet generation rate of the overall sensor network. Because an attacker may have mobility and can change its identity to spoof the remote sink, the third method is suggested in DADS. In particular, when the number of data packets arriving at the remote sink exceeds an acceptable level, the remote sink broadcasts an alarm message, called *defend against spam (DAS)*, to the network.

The basic idea of DADS is to quarantine a spam attacker by its one-hop neighbors. When a sensor node u receives a DAS message, it starts a timer t_q and only allows relay of authenticated messages before the timer expires. If node u receives an unauthenticated message from a neighbor, it asks the neighbor to resend an authenticated message. If the latter fails in authentication, node u determines that it is inside a quarantine region and will not relay any data packet unless it is successfully authenticated. In addition, node u will transmit its own messages with authentication. Figure 1.5 gives an example. Nodes a, b, c, and d are inside a quarantine region because they detect that node M fails in authentication. Then they will relay and transmit only authenticated messages. All other nodes, except node e, can still transmit unauthenticated messages to the sink. Note that node e has to send authenticated messages because node c only accepts such packets.

DADS uses the *hash-based message authentication code (HMAC)* [23], which utilizes a cryptographic one-way hash function such as MD5 [34], for message authentication. To save the authentication overhead, when a sensor node inside a quarantined region does not detect any unsuccessful authentication attempt during a period of time t_q, it switches back to the normal mode to cancel the quarantine region.

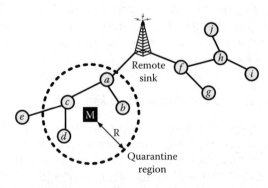

Figure 1.5   An example of DADS, where *R* is the transmission range of the malicious node *M*.

1.6   Conclusions

Communications in ad hoc and sensor networks rely heavily upon multi-hop transmission. However, most routing protocols do not address the security issue and thus trust all nodes in the network. This provides many chances for attackers to disrupt the network. This motivates many researchers to find out possible attacks and develop their countermeasures. This chapter provides a comprehensive survey of current research in attacks and defenses of routing mechanisms in ad hoc and sensor networks. Various representative attacks and their defense schemes are discussed in the chapter.

References

1. I.F. Akyildiz, W. Su, Y. Sankarasubramaniam, and E. Cayirci, A survey on sensor networks, *IEEE Communications Magazine*, Vol. 40 No. 8 (2002), pp. 102–114.
2. I. Avramopoulos, H. Kobayashi, R. Wang, and A. Krishnamurthy, Highly secure and efficient routing, *IEEE INFOCOM*, Vol. 1 (2004), pp. 197–208.
3. S. Basagni, I. Chlamtac, V.R. Syrotiuk, and B.A. Woodward, A distance routing effect algorithm for mobility (DREAM), *ACM/IEEE International Conference on Mobile Computing and Networking*, Vol. 1 (1998), pp. 76–84.
4. B. Bellur and R.G. Ogier, A reliable, efficient topology broadcast protocol for dynamic networks, *IEEE INFOCOM*, Vol. 1 (1999), pp. 178–186.
5. H. Chan and A. Perrig, Security and privacy in sensor networks, *Computer*, Vol. 36 No. 10 (2003), pp. 103–105.
6. H. Chan, A. Perrig, and D. Song, Random key predistribution schemes for sensor networks, *IEEE Symposium on Security and Privacy*, Vol. 1 (2003), pp. 197–213.
7. J. Chen, P. Druschel, and D. Subramanian, An efficient multipath forwarding method, *IEEE INFOCOM*, Vol. 1 (1998), pp. 1418–1425.
8. M.S. Corson and A. Ephremides, A distributed routing algorithm for mobile wireless networks, *Wireless Networks*, Vol. 1 No.1 (1995), pp. 61–81.
9. B.J. Culpepper and H.C. Tseng, Sinkhole intrusion indicators in DSR MANETs, *International Conference on Broadband Networks*, Vol. 1 (2004), pp. 681–688.
10. H. Deng, W. Li, and D.P. Agrawal, Routing security in wireless ad hoc networks, *IEEE Communications Magazine*, Vol. 40 No. 10 (2002), pp. 70–75.
11. L. Eschenauer and V.D. Gligor, A key-management scheme for distributed sensor networks, *ACM Conference on Computer and Communications Security*, Vol. 1 (2002), pp. 41–47.
12. X. Hong, K. Xu, and M. Gerla, Scalable routing protocols for mobile ad hoc networks, *IEEE Network*, Vol. 16 No. 4 (2002), pp. 11–21.
13. Y.C. Hu, D.B. Johnson, and A. Perrig, SEAD: Secure efficient distance vector routing for mobile wireless ad hoc networks, *IEEE Workshop on Mobile Computing Systems and Applications*, Vol. 1 (2002), pp. 3–13.

14. Y.C. Hu, A. Perrig, and D.B. Johnson, Ariadne: a secure on-demand routing protocol for ad hoc networks, *ACM/IEEE International Conference on Mobile Computing and Networking*, Vol. 1 (2002), pp. 12–23.

15. Y.C. Hu, A. Perrig, and D.B. Johnson, Packet leashes: a defense against wormhole attacks in wireless networks, *IEEE INFOCOM*, Vol. 1 (2003), pp. 1976–1986.

16. Y.C. Hu, A. Perrig, and D.B. Johnson, Rushing attacks and defense in wireless ad hoc network routing protocols, *ACM Workshop on Wireless Security*, Vol. 1 (2003), pp. 30–40.

17. K. Ishida, Y. Kakuda, and T. Kikuno, A routing protocol for finding two node-disjoint paths in computer networks, *International Conference on Network Protocols*, Vol. 1 (1995), pp. 340–347.

18. P. Jacquet, P. Mühlethaler, T. Clausen, A. Laouiti, A. Qayyum, and L. Viennot, Optimized link state routing protocol for ad hoc networks, *IEEE International Multi Topic Conference*, Vol. 1 (2001), pp. 62–68.

19. D.B. Johnson and D.A. Malts, Dynamic source routing in ad hoc wireless networks, *Mobile Computing*, edited by T. Imielinski and H. Korth, Kluwer Academic Publishers, 1996, pp. 153–181.

20. C. Karlof and D. Wagner, Secure routing in wireless sensor networks: attacks and countermeasures, *IEEE International Workshop on Sensor Network Protocols and Applications*, Vol. 1 (2003), pp. 113–127.

21. B. Karp and H.T. Kung, GPSR: greedy perimeter stateless routing for wireless networks, *ACM/IEEE International Conference on Mobile Computing and Networking*, Vol. 1 (2000), pp. 243–254.

22. Y.B. Ko and N.H. Vaidya, Location-aided routing (LAR) in mobile ad hoc networks, *Wireless Networks*, Vol. 6 No. 4 (2000), pp. 307–321.

23. H. Krawczyk, M. Bellare, and R. Canetti, RFC 2104 — HMAC: Keyed-Hashing for Message Authentication, 1997.

24. W.H. Liao, Y.C. Tseng, and J.P. Sheu, GRID: a fully location-aware routing protocol for mobile ad hoc networks, *Telecommunication Systems*, Vol. 18 No. 1 (2001), pp. 37–60.

25. H. Luo, P. Zerfos, J. Kong, S. Lu, and L. Zhang, Self-securing ad hoc wireless networks, *International Symposium on Computers and Communications*, Vol. 1 (2002), pp. 567–574.

26. S. Marti, T.J. Giuli, K. Lai, and M. Baker, Mitigating routing misbehavior in mobile ad hoc networks, *ACM/IEEE International Conference on Mobile Computing and Networking*, Vol. 1 (2000), pp. 255–265.

27. A. Mishra, K. Nadkarni, and A. Patcha, Intrusion detection in wireless ad hoc networks, *IEEE Wireless Communications*, Vol. 11 No. 1 (2004), pp. 48–60.

28. J.C. Navas and T. Imielinski, Geocast — geographic addressing and routing, *ACM/IEEE International Conference on Mobile Computing and Networking*, Vol. 1 (1997), pp. 66–76.

29. J. Newsome, E. Shi, D. Song, and A. Perrig, The Sybil attack in sensor networks: analysis & defenses, *International Symposium on Information Processing in Sensor Networks*, Vol. 1 (2004), pp. 259–268.

30. V.D. Park and M.S. Corson, A highly adaptive distributed routing algorithm for mobile wireless networks, *IEEE INFOCOM*, Vol. 1 (1997), pp. 1405–1413.

31. G. Pei, M. Gerla, and T.W. Chen, Fisheye state routing: a routing scheme for ad hoc wireless networks, *IEEE International Conference on Communications*, Vol. 1 (2000), pp. 18–22.
32. C.E. Perkins and P. Bhagwat, Highly dynamic destination-sequenced distance-vector routing (DSDV) for mobile computers, *ACM Conference on Communications Architectures, Protocols and Applications*, Vol. 1 (1994), pp. 234–244.
33. C.E. Perkins and E.M. Royer, Ad-hoc on-demand distance vector routing, *IEEE Workshop on Mobile Computing Systems and Applications*, Vol. 1 (1999), pp. 90–100.
34. R.L. Rivest, RFC 1321 — The MD5 Message-Digest Algorithm, 1992.
35. S. Sancak, E. Cayirci, V. Coskun, and A. Levi, Sensor wars: detecting and defending against spam attacks in wireless sensor networks, *IEEE International Conference on Communications*, Vol. 1 (2004), pp. 3668–3672.
36. C.K. Toh, Associativity-based routing for ad hoc mobile networks, *Wireless Personal Communications*, Vol. 4 No. 2 (1997), pp. 103–139.
37. S. Zhu, S. Setia, and S. Jajodia, LEAP: efficient security mechanisms for large-scale distributed sensor networks, *ACM Conference on Computer and Communications Security*, Vol. 1 (2003), pp. 62–72.

Chapter 2

MAC Layer Attacks in 802.15.4 Sensor Networks

Vojislav B. Mišić, Jun Fung, and Jelena Mišić

Contents

2.1   Abstract

In this chapter, we consider the networks compliant with the recent IEEE 802.15.4 standard and describe a number of possible attacks at the MAC layer. Several of these attacks can be easily launched with devices that are fully compliant with the 802.15.4 standard, and we show that such attacks can introduce serious disruption. We also discuss some remedial measures that may help defend against those attacks, or at least alleviate their impact on the performance of the network.

2.2 Introduction

Wireless sensor networks pose many challenges with respect to security [2,14,16]. From the networking perspective, security threats may occur at different layers of the ISO/OSI model [2]:

- **Routing layer** attacks include spoofed, altered, or replayed routing information spread by an adversary, selective forwarding of packets, sinkhole attacks that attract traffic from a specific area to a compromised node (or nodes), Sybil attacks in which a compromised node assumes many identities, acknowledgment spoofing, injecting corrupted packets, neglecting routing information, or forward messages along wrong paths [6,12].
- **MAC layer** attacks typically focus on disrupting channel access for regular nodes, thus disrupting the information flow both to and from the sensor node; this leads to a DoS condition at the MAC layer [16]. Security at the MAC layer has been mostly studied in the context of 802.11 MAC layer [4,8,12,17] but sometimes also in the more general context of different types of attacks [1,11,16].
- Finally, **physical layer** (jamming) attacks consist of the attacker sending signals that disrupt the information flow through radio frequency interference. Jamming at the MAC level may be accomplished by sending large-size packets with useless information.

Recently, the IEEE adopted the 802.15.4 standard for low-rate wireless personal area networks (WPANs) [5,7]. As 802.15.4-compliant WPANs use small, cheap, energy-efficient devices operating on battery power that require little infrastructure to operate, or none at all, they appear particularly well suited for building wireless sensor networks [5]. Hence, all aspects of 802.15.4 network operation and performance — security included — should be investigated and thoroughly analyzed.

In the discussions that follow, we identify a number of security threats that might affect the operation of 802.15.4 networks. Because the IEEE Std 802.15.4 [7] defines only the physical (PHY) and medium access layers (MAC) of the low-rate WPAN, we focus on the MAC layer attacks, with particular emphasis on DoS attacks [16].

The chapter is organized as follows. In Section 2.3, we describe the operation of an IEEE 802.15.4-compliant sensor network, including the security provisions prescribed by the standard. Section 2.4 lists and briefly describes two classes of possible attacks at the MAC level, distinguished by compliance with the operation of the MAC protocol or lack thereof. Section 2.5 discusses the impact of some of those attacks, while Section 2.6 describes the manner in which some of those attacks could be prevented or their effects minimized. Section 2.7 concludes the chapter and outlines some directions for future work.

2.3 Operation of the 802.15.4 MAC

In an IEEE 802.15.4-compliant WPAN, a central controller device (commonly referred to as the PAN coordinator) builds a WPAN with other devices within a small physical space known as the personal operating space. Two topologies are supported: (1) in the star topology network, all communications, even those between the devices themselves, must go through the PAN coordinator; (2) in the peer-to-peer topology, the devices can communicate with one another directly — as long as they are within the physical range — but the PAN coordinator must be present nevertheless. The standard also defines two channel access mechanisms, depending on whether a beacon frame (which is sent periodically by the PAN coordinator) is used to synchronize communications or not. Beacon-enabled networks use slotted carrier sense multiple access mechanism with collision avoidance (CSMA-CA), while the non-beacon-enabled networks use simpler, unslotted CSMA-CA.

In beacon-enabled networks, the channel time is divided into superframes that are bounded by beacon transmissions from the coordinator [7]. All communications in the cluster take place during the active portion of the superframe, the duration of which is referred to as the superframe duration *SD*. During the (optional) inactive portion, nodes may enter a low power mode, or engage in other activities at will.

The active portion of each superframe is divided into equally sized slots; the beacon transmission commences at the beginning of slot 0, and the contention access period (CAP) of the active portion starts immediately after the beacon. Slots are further subdivided into backoff periods, the basic time units of the MAC protocol to which all transmissions must be synchronized. The actual duration of the backoff period depends on the frequency band in which the 802.15.4 WPAN is operating; in the highest, 2.4-GHz band, the maximum data rate is 250 kbps [7].

A part of the active portion of the superframe may be reserved by the PAN coordinator for dedicated access by some devices; this part is referred to as the contention-free period (CFP), while the slots within are referred to as the guaranteed time slots (GTS).

2.3.1 The CSMA-CA Algorithm

During the CAP, individual nodes access the channel using the CSMA-CA algorithm, the operation of which is schematically shown in Figure 2.1.

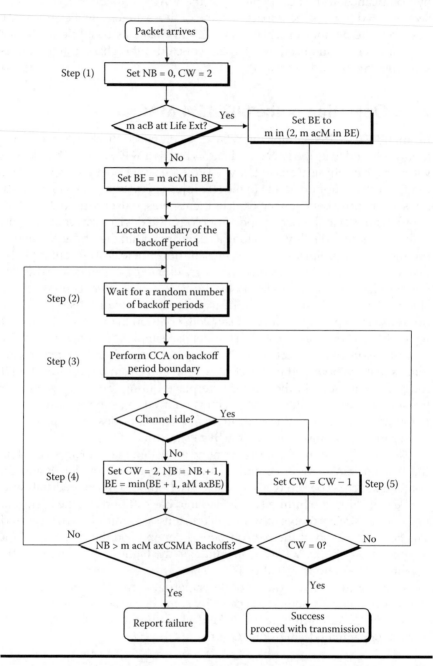

Figure 2.1 Operation of the slotted CSMA-CA MAC algorithm in the beacon-enabled mode. (Adapted from [7].)

The algorithm begins by initializing *NB* to zero and *CW* to 2; the variable $NB = 0..$ *macMaxCSMABackoff* − 1 represents the index of the backoff attempt, while the variable *CW* = 0, 1, 2 represents the index of the Clear Channel Assessment (CCA) phase counter.

If the device operates on battery power, as indicated by the attribute *macBattLifeExt*, the parameter *BE* (the backoff exponent that is used to calculate the number of backoff periods before the node device attempts to assess the channel) is set to 2 or to the constant *macMinBE*, whichever is less; otherwise, it is set to *macMinBE* (the default value of which is 3). The algorithm then locates the boundary of the next backoff period; as mentioned above, all operations must be synchronized to backoff time units.

In step (2), the algorithm generates a random waiting time *k* in the range $0.. 2^{BE} − 1$ backoff periods. The value of *k* is then decremented at the boundary of each backoff period. Note that the counter will be frozen during the inactive portion of the beacon interval, and the countdown will resume when the next superframe begins.

When this counter becomes zero, the device must make sure the medium is clear before attempting to transmit a frame. This is done by listening to the channel to make sure no device is currently transmitting. This procedure, referred to as Clear Channel Assessment (CCA), must be done in two successive backoff periods, as shown by steps (3) and (5) in Figure 2.1. If both CCAs report that the channel is idle, packet transmission may begin.

If the channel is busy at the first CCA, the values of *i* and *BE* are increased by one, while *c* is reset to 2, and another random wait is initiated; this is step (4) in the flowchart. In this case, when the number of retries is below or equal to *macMaxCSMABackoffs* (the default value of which is 5), the algorithm returns to step (2); otherwise it terminates with a channel access failure status; it is up to the higher protocol layers to decide whether to re-attempt the transmission as a new packet or not. However, if the channel is found busy at the second CCA, the algorithm simply repeats the two CCAs starting from step (3).

Before undertaking step (3), the algorithm checks whether the remaining time within the CAP area of the current superframe is sufficient to accommodate the CCAs, the data frame, the proper interframe spacing, and the acknowledgment. If this is the case, the algorithm proceeds with step (3); otherwise it will simply pause until the next superframe, and resume step (3) immediately after the beacon frame.

2.3.2 On Uplink and Downlink Communication

According to the 802.15.4 standard, uplink data transfers from a node to the coordinator are synchronized with the beacon, in the sense that both the original transmission and the subsequent acknowledgment must occur

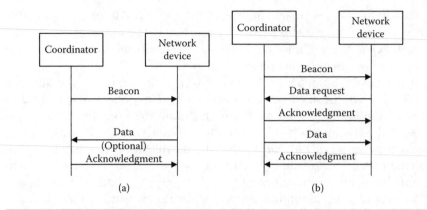

Figure 2.2 Uplink and downlink data transfers in beacon-enabled PAN: (a) uplink transmission and (b) downlink transmission.

within the active portion of the same superframe, as shown in Figure 2.2a. Uplink transmissions always use the CSMA-CA mechanism outlined above.

Data transfers in the downlink direction, from the coordinator to a node, are more complex, as they must first be announced by the coordinator. In this case, the beacon frame will contain the list of nodes that have pending downlink packets, as shown in Figure 2.2b. When the node learns there is a data packet to be received, it transmits a MAC command requesting the data. The coordinator acknowledges the successful reception of the request by transmitting an acknowledgment. After receiving the acknowledgment, the node listens for the actual data packet for the period of *aMaxFrameResponseTime*, during which the coordinator must send the data frame.

According to the standard, it is allowed to send the data frame "piggybacked" after the request acknowledgment packet, that is, without using CSMA-CA. However, two conditions must be fulfilled: (1) the coordinator must be able to commence the transmission of the data packet between *aTurnaroundTime* and *aTurnaroundTime* + *aUnitBackoffPeriod*, and (2) there must be sufficient time in the CAP for the message, appropriate interframe spacing, and acknowledgment; if either of these is not possible, the data frame must be sent using the CSMA-CA mechanism [7]. While the first condition depends on the implementation platform, the second depends on the actual traffic, and *some* data frames may have to be sent using CSMA-CA.

Note that acknowledgments are sent only if explicitly requested by the transmitter; the acknowledgment must be received within a prescribed time interval; otherwise, the entire procedure (starting from the announcement in the beacon frame) must be repeated.

2.3.3   Security Services and Suites

The 802.15.4 standard specifies several security suites, which consist of a "set of operations to perform on MAC frames that provide security services" [7]. Specified security services include the following:

- Any device can maintain an Access Control List (ACL) — a list of trusted devices from which the device wishes to receive data; this mechanism is intended to filter out unauthorized communications.
- Data Encryption service helps a device encrypt a MAC frame payload using the key shared between two peers, or among a group of peers. If the key is to be shared between two peers, it is stored with each entry in the ACL list; otherwise, the key is stored as the default key. (The MAC layer provides the symmetric encryption security systems using an application-provided key, or keys.) Thus, the device can make sure that the data cannot be read by devices that do not possess the corresponding key. However, device addresses are always transmitted in the clear (i.e., unencrypted), which makes attacks that rely on device identity somewhat easier to launch.
- Frame Integrity service ensures that a frame cannot be modified by a receiver device that does not share a key with the sender, by appending a message integrity code (MIC) generated from blocks of encrypted message text.
- Finally, Sequential Freshness uses the frame counter and key sequence counter to ensure the freshness of the incoming frame and guard against replay attacks.

The services listed above are typically implemented in hardware for performance reasons, and their use is optional. A device can choose to operate in unsecured mode, secured mode, or ACL mode. In unsecure mode, none of the services mentioned above are available. In secured mode, the device can use one of the security suites supported by the standard [7], all of which use the Data Encryption service explained above.

A device operating in ACL mode can maintain a list of trusted devices from which it expects to receive packets, but the only security service available is access control service, which enables the receiver to filter received frames according to the source address listed in the frame. However, because no encryption is used, it is not possible to authenticate the true source of the data packet, or to ascertain that the packet payload has not been modified in any way.

We note that the procedures for key management, device authentication, and freshness protection are not specified by the standard; they are left to be implemented by the applications running on 802.15.4 devices.

2.4   MAC Layer Attacks

Attacks can be broadly classified into two categories, depending on whether the attacker follows the rules of the 802.15.4 MAC protocol, either fully or only to a certain extent, or not. While the attacks from the latter category are potentially more dangerous, defense against them is much more difficult, as might be expected; in this case, the attacker can use a separate 802.15.4-compliant device, possibly modified to lose adherence to the MAC protocol. Alternatively, an existing 802.15.4 device can be captured and subverted so as to be used for malicious purposes.

We note that an adversary with appropriate resources might develop and use dedicated hardware that is compatible but not compliant with the 802.15.4 standard. The discussion of such attacks is beyond the scope of this chapter.

2.4.1   *Attacks that Follow the MAC Protocol*

It may come as a surprise that a number of attacks can be conducted by an adversary that follows the IEEE 802.15.4 slotted CSMA-CA protocol to the letter. All of those attacks can be conducted by an adversary that acts as a legitimate member of the PAN.

A simple but not very efficient attack against network availability is to flood the network by simply transmitting a large number of packets. Packets should be large in size, perhaps the largest size allowed by the standard. In this manner, an adversary can degrade the network performance and drastically reduce throughput; our previous work indicates that the performance of an 802.15.4 network can be seriously affected by high packet arrival rates or by nodes operating in saturation regime [9,10].

An adversary can target different destination devices with unnecessary packets, possibly in other PANs, regardless of whether the destination PAN or device actually exists or not. If the goal of the attack is the depletion of the power source for a specific node (and the PAN coordinator), all injected packets can target that node. Because the downlink packets have to be explicitly requested from the PAN coordinator, this will keep both the PAN coordinator and the chosen destination device busy and eventually exhaust their respective power sources.

A malicious node can simply pretend to run in battery life extension mode by setting the *aMacBattLifeExt* variable to true. In that case, the CSMA-CA algorithm will choose the initial value of the backoff exponent as 2 instead of 3, as explained in Section 2.3.1, and the random number for the backoff countdown will be in the range 0 .. 3 rather than in the range 0 .. 7. Shorter backoff countdown means that the probability of accessing the medium is much higher than for a regular node. On top of that, a regular node would have to wait for the malicious one to finish its transmission, and waste power in the process.

Note also that the node that succeeds in getting access to the medium will not increase its backoff exponent for the next transmission, while the unsuccessful one will increase it by one. Therefore, if the first attempt succeeded, the second one is even more likely to do so, which again clearly favors malicious nodes.

It should be noted that power consumption during packet reception is about one half to two thirds of the corresponding power consumption required for packet transmission [5]. Therefore, while transmitting does consume lots of energy (relatively speaking), receiving is not terribly efficient either, and the best way to conserve power is to turn off the radio subsystem whenever possible.

2.4.2 Attacks that Use a Modified MAC Protocol

The attacks mentioned above can be conducted using a completely functional 802.15.4 device that follows the protocol to the letter — it suffices for the malicious node to control the application that executes on the sensor device. However, a number of additional attacks can be launched by simply modifying or disregarding certain features of the protocol. This can be accomplished either through dedicated hardware or by controlling an otherwise fully compliant 802.15.4 hardware device, as follows.

We have already mentioned the possibility of decreasing the backoff exponent and shortening the random backoff countdown by falsely reporting that battery life extension is enabled. Similar effects can be achieved by *not* incrementing the backoff exponent after an unsuccessful transmission attempt.

The random number generator can be modified to give preference to shorter backoff countdowns. Again, this allows the malicious node to capture the channel in a disproportionately high number of cases, and gives it an unfair advantage over regular nodes.

The number of required CCA attempts can be reduced to one instead of two, which would give the malicious node an unfair advantage over the regular nodes.

The CCA check can be omitted altogether, in which case the node will start transmitting immediately after finishing the random backoff countdown. Even worse, the node can omit the random backoff countdown itself. In this manner, the malicious node can transmit its packets more often than a regular one. While not all of the messages will be sent successfully — there will be collisions in many cases — the malicious node probably does not even care, as long as the transmissions from regular nodes end up garbled and thus must be repeated. Moreover, some of the attacker's transmissions may collide with acknowledgments. Again, this wastes the power of the devices affected, and also the bandwidth of the entire network.

In case the acknowledgment is requested by the data frame or the beacon frame, a malicious node may simply refuse to send it. The PAN coordinator will retry transmission (up to a maximum of *aMaxFrameRetries*) and thus waste power and bandwidth.

2.5   Impact of Attacks that Follow the MAC Protocol

To assess the impact of attacks, we have built the simulator of an IEEE 802.15.4-compliant sensor cluster at the MAC level, using the object-oriented Petri Net-based simulation engine Artifex by RSoft Design, Inc. [13]. The cluster operates in the ISM band at 2.4 GHz, in a star configuration with the PAN coordinator acting as the network sink, a total of 20 regular nodes, and one or two attacker nodes. Slotted, beacon-enabled communication with CSMA-CA mechanism described in Section 2.3.1 is used for all communications, for reasons outlined in Section 2.3.2. Regular nodes generate Poisson distributed traffic with packets of three backoff periods, which corresponds to the payload of 30 bytes, with the average arrival rate of 120 packets per minute.

The following attack scenarios from Sections 2.4.1 and 2.4.2 were considered:

- The attacker node follows the MAC protocol to the letter but generates spurious traffic with varying packet size and arrival rate. This is the simplest form of attack: flooding the cluster with useless packets.
- The attacker node tries to gain unfair advantage by operating with the *aMacBattLifeExt* variable set; this reduces the random backoff exponent and reduces the range for random backoff countdown. Packet length and arrival rate are variable.
- The attacker node uses one CCA check instead of two required by the standard; it also has the *aMacBattLifeExt* variable set and uses variable packet length and arrival rate.

In each scenario, we have measured the probability that the regular node will succeed in transmitting its packets, as well as the mean packet delay for those packets. In this manner, we can obtain a quantitative measure of the disruption caused by the attacker node, or nodes. Note that the traffic and network parameter values are chosen so that the cluster operates in non-saturation mode.

Figure 2.3 shows the success probability for the regular nodes in the three scenarios outlined above, in clusters with one and two attacker nodes, respectively. As can be seen, the success probability rapidly decreases with increasing length and injection rate of the packets generated by the attacker node(s). As can be expected, two attacker nodes generate more disruption

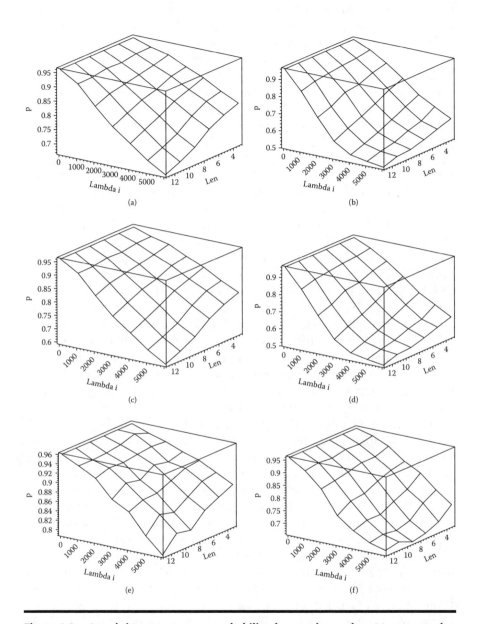

Figure 2.3   Attack impact: success probability for regular nodes: (a) one attacker node; (b) two attacker nodes; (c) one attacker node, *aMacBattLifeExt* set; (d) two attacker nodes, *aMacBattLifeExt* set; (e) one attacker node, aMacBattLifeExt variable set, and a single CCA; and (f) two attacker nodes, aMacBattLifeExt variable set, and a single CCA.

than one. This is particularly pronounced in the second and third scenarios described above. In the worst case, Figures 2.3b and 2.3d, the success probability for regular transmissions drops to about 0.5, as opposed to the "normal" value of about 0.95. That is, the energy expenditure per packet will be twice the value needed in normal operation, which means that the cluster lifetime — if operating on battery power — will be cut in half. Note that those values were obtained in a cluster with 20 regular nodes that is attacked by two nodes only — and those nodes still follow the MAC protocol to the letter!

Somewhat unexpectedly, the use of one CCA check instead of two does not decrease the success probability for regular nodes — actually, it improves it instead. Namely, collisions happen when two or more nodes finish their random backoff countdown at the same time, check that the medium is idle during two CCAs, and then start transmissions simultaneously. By skipping the second CCA, the attacker node is able to gain access to the medium faster than a regular node. In such cases, however, the regular node will see the medium busy in *its* second CCA and it will restart the CSMA-CA algorithm with the backoff exponent increased by one. Consequently, the probability of collision between transmissions from a regular node and an attacker node is reduced. The decrease in collision probability means that the success probability increases, as witnessed by measurements in Figures 2.3e and 2.3f. Note that collisions still occur when the attacker node finishes its backoff countdown in the backoff period in which a regular node performs its first CCA; and the probability of collisions between two regular nodes is not affected at all.

The argument described above is illustrated through the diagrams of success probabilities for the first and second CCA for regular nodes shown in Figure 2.4. For brevity, we show only the case with two attacker nodes. As can be seen from Figures 2.4a, 2.4c, and 2.4e, the probability that the first CCA is successful does deteriorate with increased packet arrival rate and packet length. However, it is not particularly affected by the introduction of reduced backoff exponent or by the attacker nodes skipping one CCA; the differences between successive diagrams is rather small.

Similarly, the success probability for the second CCA does not change much from the case where attackers follow the MAC protocol to the letter, Figure 2.4b, to the one in which they use the reduced backoff exponent, Figure 2.4d. In fact, the second CCA even shows some recovery at high packet arrival rates and large packet length; note, however, that this is the *second* CCA, and only the transmissions that have successfully passed the first one get to this point.

However, when the attacker nodes skip the second CCA themselves, the situation changes. As can be seen from Figure 2.4f, the probability that a regular node will succeed in its second CCA is radically reduced, compared to the diagrams above. (Note that success in both CCAs does not guarantee

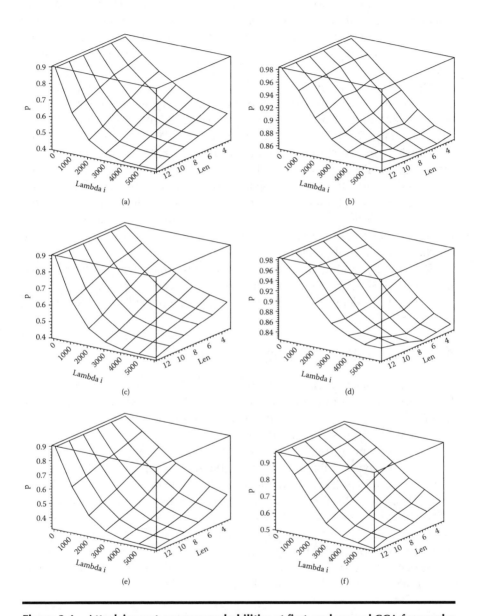

Figure 2.4   Attack impact: success probabilities at first and second CCA for regular nodes (two attacker nodes): (a) first CCA; (b) second CCA; (c) first CCA, attackers have *aMacBattLifeExt* set; (d) second CCA, attackers have *aMacBattLifeExt* set; (e) first CCA, attackers have *aMacBattLifeExt* set and use a single CCA; and (f) second CCA, attackers have *aMacBattLifeExt* set and use a single CCA.

a successful transmission, as collisions can still occur.) Therefore, a large fraction of transmissions from regular nodes will fail the second CCA and will have to wait for at least another pass of the CSMA-CA algorithm. The net result is that the regular packets will experience much larger transmission delays.

This behavior is obvious from the diagrams in Figure 2.5, which show mean end-to-end delays for the packets sent by regular nodes in the three scenarios outlined above. All delays are expressed in units of backoff periods, which last for 0.32 ms if the network operates in the 2.4 GHz ISM band. We consider sensor clusters with one and two attacker nodes, respectively. Delays generally increase when the length or injection rate of the packets generated by the attacker node(s) increases. When two attacker nodes are present, delays are about 70 to 80 percent larger than in the case with only one attacker node, as can be expected. Furthermore, the impact of different attack mechanisms on packet delay is cumulative — that is, each attack mechanism makes the delays longer.

We end by noting that some applications may actually tolerate the increased delays, provided the increase in success probability suffices; in other cases, delay is a critical factor and increased delays cannot be tolerated.

2.6   Defending the 802.15.4 PAN

While the attacks listed in the previous subsection may indeed pose formidable risks to normal operation of an 802.15.4 WPAN, it should be noted that they are probably not very cost-effective to launch. Because individual 802.15.4 sensor nodes are small, low-power, low-cost devices, the development of dedicated compatible-but-not-compliant devices with modified behavior is likely to be prohibitively expensive — the potential attacker would probably find the use of simple devices for jamming at the PHY layer to be much more attractive. Let us consider just the 2450-MHz (the so-called ISM band), with the raw data rate of 250 kbps, which is already used by other wireless LAN/PAN standards such as 802.11b and 802.15.1 (Bluetooth), and high interference may be expected. From the specifications of the 802.15.4 standard [7], the processing gain is only around 8, and the Bit Error Rate (BER) is given by

$$BER = Q\left(\sqrt{\frac{E_b}{N_0}}\right)$$

where $Q(u) \approx e^{-u^2/2}(\sqrt{2\pi}\,u)$, $u \gg 1$ [3]. Therefore, in an interference-free environment, we should expect the BER values slightly below 10^{-4}. As the packet error rate is $PER = 1 - (1 - BER)^X$, where X is the total packet length (expressed in bits) including MAC and physical layer headers, the

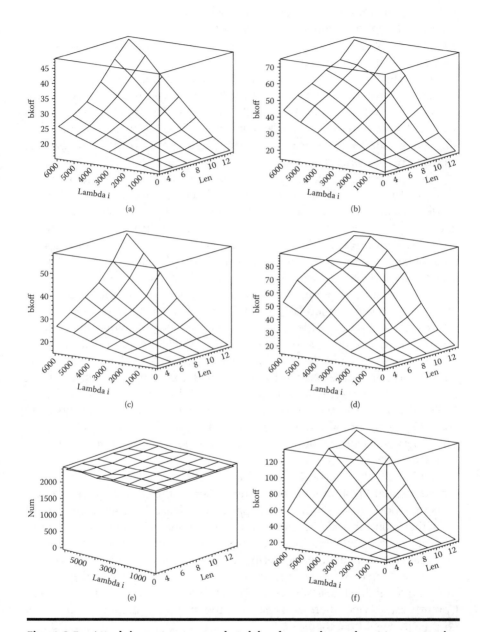

Figure 2.5 **Attack impact: mean packet delay for regular nodes: (a) one attacker node; (b) two attacker nodes; (c) one attacker node, *aMacBattLifeExt* set; (d) two attacker nodes, *aMacBattLifeExt* set; (e) one attacker, *aMacBattLifeExt* variable set, and a single CCA; and (f) two attacker nodes *aMacBattLifeExt* variable set, and a single CCA.**

probability that a given packet (data or acknowledgment) is much higher: Section 6.1.6 of the standard states that a PER of 1% is expected on packets with length of 20 octets [7]. The use of shorter packets will increase the resiliency of the transmission.

However, in the presence of interference and noise in the ISM band, higher BER values, and much higher PER values, can be expected; for example, at BER around 10^{-3}, the PER for a 20-octet packet increases to around 15%. Consequently, an attacker who chooses to jam the transmissions in an 802.15.4 sensor network can cause quite a lot of damage with modest energy expenditure.

It is worth noting that the provisions of Section 6.7.9 of the standard allow 802.15.4 devices to perform CCA checks in one of three different modes: by energy only (mode 1), by carrier sense only (mode 2), or by energy and carrier sense (mode 3) [7]. A legitimate node that uses mode 1 can experience CCA failures in a high interference environment, without an attacker specifically using 802.15.4 modulated signal. Obviously, CCA mode 3 should be used for highest resilience.

In terms of encryption, the obvious weakness of the standard is that the encryption is applied to packet/frame payload only, but not to the information in packet headers; this holds not only for ordinary data packets, but also for data request and beacon frames. While the address list field — the list of addresses of devices that have pending downlink packets — is considered part of the beacon frame payload according to the standard, it is not encrypted even when the security is enabled for the beacon frame. It would not be too big a change of the standard to encrypt not just the packet payload, but also the destination address and the address list fields. This simple measure would make some of the attacks much more difficult to accomplish.

Replay attacks can be identified by the PAN coordinator — provided the Sequential Freshness service is used. While this would not prevent a malicious node from sending such packets, at least the coordinator could filter them out and avoid any further processing. However, this feature may not be very efficient in terms of memory required. Namely, each device must maintain a table for storing the monotonically increasing counter values associated with message streams sent by each entry in the ACL. If the Sequential Freshness service is enabled in conjunction with the Access Control List service, there is a risk that the available memory of the PAN coordinator will be exhausted.

While the standard does not prescribe, or indeed even recommend, any particular key management scheme, the overall effectiveness of the security services provided by the standard depends very much on the choice of a suitable scheme [15].

Intrusion detection techniques could help identify malicious nodes that might be trying to disrupt the normal operation of the PAN. By analyzing the traffic patterns, the PAN coordinator may become aware of the activity of such nodes, so that appropriate measures can be taken to minimize the disruption. Suspicious activities include amounts of traffic well above the average, traffic intensity that increases over time, possibly in an abrupt fashion, and sending packets to many destinations, possibly in different PANs. In more critical applications, devices with substantially higher computational capabilities (and operating on mains power, rather than battery power) could analyze the activities of individual nodes at the MAC level and identify potential intruder(s).

We note that sensing applications often involve operation with very low duty cycle of individual nodes; this makes intrusion detection comparatively easier to accomplish, and does not help the attackers who might be eager to achieve their objectives.

The standard does not provide for periodic checking of the presence or integrity of individual devices. However, a sensing application might establish such checks on its own. Simple time-out counters, one per each associated device, would enable the coordinator to check for their continued presence; in addition, a simple challenge/response scheme could allow the coordinator to verify their integrity as well.

Of course, it is impossible to physically isolate an unwanted device in a wireless network; but at least the application should be made aware of the presence of such devices so that their impact on the normal operation of the network can be minimized.

2.7 Conclusions

We have investigated the security of IEEE 802.15.4 sensor networks at the MAC level. A number of vulnerabilities have been identified, and some simple attack mechanisms that exploit those vulnerabilities have been described and analyzed. More work is needed to address the possible defenses against those and other attack mechanisms.

References

1. J. Bellardo and S. Savage. 802.11 denial-of-service attacks: real vulnerabilities and practical solutions. In *Proc. 12th USENIX Security Symposium*, Washington, D.C., August 2003.
2. H. Chan and A. Perrig. Security and privacy in sensor networks. *IEEE Computer*, 36(10):103–105, October 2003.
3. V.K. Garg, K. Smolik, and J.E. Wilkes. *Applications of CDMA in Wireless/ Personal Communications*. Prentice Hall, Upper Saddle River, NJ, 1998.
4. V. Gupta, S. Krishnamurthy, and M. Faloutsos. Denial of service attacks at the MAC layer in wireless ad hoc networks. In *Proc. MILCOM 2002*, Anaheim, CA, October 2002.

5. J.A. Gutiérrez, E.H. Callaway, Jr., and R.L. Barrett, Jr. *Low-Rate Wireless Personal Area Networks.* IEEE Press, New York, 2004.
6. Y.-C. Hu and A. Perrig. A survey of secure wireless ad hoc routing. *IEEE Security & Privacy Magazine*, 2(3):28–39, May–June 2004.
7. Standard for part 15.4: Wireless MAC and PHY Specifications for Low Rate WPAN. IEEE Std 802.15.4, IEEE, New York, October 2003.
8. C. Karlof, N. Sastry, and D. Wagner. TinySec: a link layer security architecture for wireless sensor networks. In *Proc. SenSys 2004*, pages 162–175, Baltimore, MD, November 2004.
9. J. Mišić and V.B. Mišić. Access delay and throughput for uplink transmissions in IEEE 802.15.4 pan. *Elsevier Computer Communications*, to appear, 2005.
10. J. Mišić, S. Shafi, and V.B. Mišić. Analysis of 802.15.4 beacon enabled PAN in saturation mode. In *Proc. SPECTS 2004*, San Jose, CA, July 2004.
11. J. Newsome, E. Shi, D. Song, and A. Perrig. The Sybil attack in sensor networks: analysis and defenses. In *Proc. IPSN 2004*, pages 259–268, Berkeley, CA, April 2004.
12. Q. Ren and Q. Liang. Secure media access control (MAC) in wireless sensor networks: intrusion detections and countermeasures. In *Proc. PIMRC 2004*, Volume 4, pages 3025–3029, Barcelona, Spain, September 2004.
13. RSoft Design, Inc. *Artifex v.4.4.2.* San Jose, CA, 2003.
14. E. Shi and A. Perrig. Designing secure sensor network. *IEEE Wireless Communications*, 11(6):38–43, December 2004.
15. W. Stallings. *Cryptography and Network Security — Principles and Practice.* Prentice Hall, Upper Saddle River, NJ, 3rd edition, 2003.
16. A.D. Wood and J.A. Stankovic. Denial of service in sensor networks. *IEEE Computer*, 35(10):54–62, October 2002.
17. Y. Zhou, D. Wu, and S.M. Nettles. Analyzing and preventing MAC-layer denial of service attacks for stock 802.11 systems. In *Proc. BroadWISE 2004*, pages 22–88, San Jose, CA, October 2004.

ENCRYPTION, AUTHENTICATION, AND WATERMARKING

II

Chapter 3

Securing Radio Frequency Identification

Yang Xiao and Swapna Shroff

Contents

3.1 Abstract

Radio frequency identification (RFID) can uniquely identify objects using wireless radio communications. An RFID system consists of three components: tag, reader, and database. A tag attached to an object is a small microchip with an integrated circuit, and it stores information about the object, including the object's unique identification. A reader interrogates tags via wireless radio channels for the objects' information and connects the database to the computer. RFID tags must be low cost so that they are limited in computation and communication capacity, and thus lead to various security and privacy issues. It is a challenging task to provide better security and privacy with low-cost RFID rags. This survey addresses various security and privacy issues that a low-cost RFID tag encounters, and discusses some security and privacy solutions proposed in the literature. We also point out some problems in these various solutions and propose some enhancements.

3.2   Introduction

Radio frequency identification (RFID) is an automatic identification technology that uniquely identifies objects or people using wireless radio communications. An RFID system consists of three components: tags (also called transponders), handheld or stationary interrogators (also called readers), and a back-end database, as shown in Figure 3.1. An RFID tag consists of a small silicon integrated microchip attached to an antenna. Data is stored on a microchip, which also performs logical operations. Memory on tags can be read-only, write-once read many, or fully rewritable. An RFID reader is basically a transmitter and receiver connected to the database in the computer. It collects the data from tags and passes them to the computer for processing. RFID tags and readers communicate via antennas using wireless radio channels.

RFID tags are further classified into active tags, semi-active tags, passive tags, and semi-passive tags. An active tag has an on-board power source such as a battery to run the microchip's circuit, can broadcast a signal to the readers, and does not require a reader to interrogate it first. A passive tag has no battery and depends on the reader's radio frequency (RF) signal

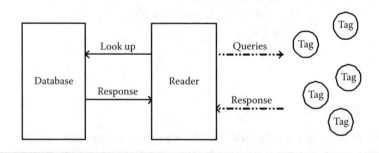

Figure 3.1   An example RFID system.

for powering its microchip to transmit. A semi-active tag has a battery for powering its antenna for transmission but depends on the reader's RF energy for powering the chip. A semi-passive tag contains a small on-board battery to power the chip but cannot power-on the antenna, and depends on the reader's energy for transmission. A tag's power limits the distance from which a reader can interrogate tags. Therefore, an active tag can be read from a greater distance than a passive tag. If its battery dies, an active tag becomes useless. An active tag in the present market costs about $0.50, and a passive tag is around $0.05 when ordering a large quantity. Active tags can be used to track costly items over long distances, and passive tags can be used to track cheap items over short distances.

Different frequency ranges are used for tags for different kinds of applications. Low-frequency (LF) tags operate at 30 to 500 kHz at close distances with a slower rate, and are cheaper and can penetrate through nonmetallic substances such as water and dirt. High-frequency (HF) tags operate at 900 MHz to 2.5 GHz with distances up to 250 feet, and can also penetrate through nonmetallic substances. Ultra-high-frequency (UHF) tags operate at 915 MHz with a longer transmission range over 10 feet and a faster transmission rate but cannot penetrate through nonmetallic substances. Microwave tags operate at 2.45 GHz for long-range access applications.

Figure 3.1 shows an example RFID system in which each object to be identified is tagged or embedded with an RFID tag. A unique identification (ID) is used to identify the object and is stored in the tag's microchip. A reader sends a query to the tag asking for the tag's ID. Using the electrical impulse of an RF signal received from the reader, the tag powers-on its onboard chip if it is a passive tag. The tag responds to the query with its ID. Readers use the tag's ID as a look-up key to query the database to obtain records associated with the tag's ID.

There are many RFID applications. A detailed survey can be found in Xiao et al. [1]. Here, we just list a few of them that are widely used:

- RFID technology has been used in supply chain management, pallet tracking, and inventory control.
- RFID systems have been used in automatic toll payment systems, in which vehicles equipped with RFID tags and readers are attached to toll lanes to scan tag data and to perform deduction from users' accounts.
- Low-frequency RFID tags are used to track movements of animals or pets or to identify IDs of them.
- RFID tags are used in libraries to track books, serve as security devices, and to be convenient to check out.
- RFID technology is used for physical access control of buildings/doors.
- RFID systems are used in airports for baggage tracking.

Both RFID tags and barcodes are intended to provide identification for objects. A barcode scanner requires line-of-sight of a barcode to read, whereas RFID tags can be read without physical access and line-of-sight, and even through materials such as paper and water. Barcode information is not unique to each item, whereas an RFID tag is unique to each item. That is, the barcode on a coke tin in Memphis might be the same as the barcode on a coke tin in New York; but in the RFID system, no two items have the same identification number; that is, the identification number stored in the tag on the coke tin in Memphis will not be present on any other coke tin anywhere in the world. Hence, an RFID system distinguishes one coke tin from the many other millions of exactly the same type. Many RFID tags can be read simultaneously, whereas barcodes cannot. Some people believe that RFID technology will replace barcode technology in the near future. Some other people believe that due to the high cost and privacy issue of RFID, both RFID tags and barcodes will coexist.

The rest of the chapter is organized as follows. Section 3.3 addresses security and privacy issues related to RFID. Section 3.4 briefly explains anti-collision protocols. Section 3.5 presents various security solutions proposed in the literature. Section 3.6 concludes the chapter.

3.3   Security and Privacy Issues

RFID tags are low-cost so that they have limited computational capacity; for example, even basic symmetric-key cryptographic operations are too expensive. Therefore, it is a challenging task to provide the required security for low-cost RFID tags.

RFID tags may pose privacy and security risks to both organizations and individuals. RFID tags can be easily traced. If tags are read freely, both personal location privacy and corporate security may be violated. RFID tags can also be cloned or duplicated. Because identification numbers stored in tags are static, an adversary can duplicate tags by obtaining the information. This attack is of particular concern when RFID tags are used for payment systems and access controls. An adversary can obtain the identification number of a tag, and also can target the back-end database to use the identification number as a lookup to get access.

Some privacy concerns are listed as follows:

- A customer is not aware of the presence of tags in purchased items.
- RFID tags can be read from a distance without the knowledge of their users.
- RFID tags embedded in clothes or purses can be easily scanned because they communicate with readers using low-frequency radios that can go through fabric, plastic, and other materials.

- In the current market, the majority of RFID tags do not require authentication to read so that they can be scanned from a longer distance using a high-gain antenna. The tag contents in one's purse may make one vulnerable to mugging if high value is known by other people.
- For example, assume that customer B purchases a pair of shoes in a store using his credit card. The store can make an association between B's name and the unique tag ID on his shoes, and the store can further link the tag ID with B's credit history.
- RFID tags also pose risks of corporate espionage; for example, a competitive manufacturer B can easily know how many products manufacturer A produces each year by scanning A's tags. B also can overwrite A's tags as well as launch a DoS attack.

3.4 Anti-collision Protocols

An RFID reader can communicate with only a single RFID tag at a time. For example, in a supermarket at an automated checkout, if a reader tries to scan the tags that are in its range, then all the tags respond with their IDs and collisions can occur. Furthermore, if two readers query the same tag at the same time, the query messages can collide. Hence, an anti-collision protocol is needed. The most common anti-collision protocols are tree-walking and ALOHA. Tree-walking and ALOHA can be used by the tags at 915 MHz and 13.56 MHz, respectively.

In the tree-walking protocol, shown in Figure 3.2, a reader queries a tag by sending one bit of the tag's ID at a time until the reader finds the tag it

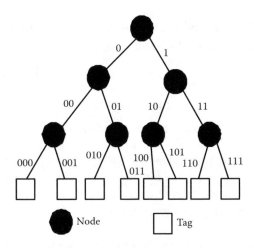

Figure 3.2 Tree-walking with tag IDs.

is looking for. In the example shown in Figure 3.2, 3-bit IDs are treated as leaves in a depth-first tree of depth 3. The root is treated as the node with depth 0. For a given node in the tree, the reader queries the tags whose ID starts with a particular prefix, and the tags that have the specified prefix reply to the query and others remain silent. If the reader asks for the tags to answer if their ID starts with 1 or 0 and if more than one tag responds to the query, the reader will ask for the tags with an ID that starts with 01 to respond and then 010.

Passive tags need energy from the reader to initiate the transmission. Therefore, any information sent from the reader to a tag is transmitted with more power and hence vulnerable to eavesdropping. The channel through which the data is transmitted from the reader to a tag is called the downlink, and the channel from a tag to the reader is called the uplink. The tree-talking protocol is vulnerable to eavesdropping because the reader broadcasts the next bit of the tag ID in the downlink, asking for a tag that has that bit to respond.

In the ALOHA protocol, when a collision is detected, the tags wait for some random amount of time and resend the IDs to the reader. This protocol is less vulnerable to eavesdropping than the tree-walking protocol, but it may waste the available bandwidth when the reader field is densely populated.

3.5 Security Solutions

This section discusses the various security solutions proposed in the literature for RFID.

3.5.1 Hash-Based and Randomized Access Controls, and Silent Tree Walking

Weis et al. [2] address security and privacy aspects of RFID such as eavesdropping, traffic analysis between a reader and a tag, spoofing, or DoS of RFID under the following assumptions:

- Tags are passive and read-only, with a few hundred bits of storage, an operating range of a few meters, and 200 to 2000 gates.
- A tag's memory is insecure, such that if the tag is exposed to physical attacks, all the data in its memory is revealed. Therefore, assume that the tag cannot be trusted to store long-term secrets such as keys.
- Assume that connections between tag readers and the back-end database are secure.
- The downlink can be monitored from a longer distance, such as 100 meters, whereas the uplink can be monitored only from a shorter range.

With above assumptions, Weis et al. [2] propose three security solutions as follows.

3.5.1.1 Hash-Based Access Control

In hash-based access control [2], tags can be locked and unlocked, and a one-way hash function is adopted. To lock a tag, the following procedure is used:

1. A metaID is calculated by the reader that is the result of a one-way hash function with an input of a random key *key* as follows: $metaID \leftarrow hash(key)$.
2. The reader stores $(metaID, key)$ in the back-end database.
3. The reader sends the metaID either through an RF channel or a physical contact channel to the tag.
4. After receiving this metaID, the tag enters into the locked state, during which period the tag responds to all queries with only its metaID.

The reader can unlock the tag in the following way:

1. The reader sends a query to the tag.
2. The tag sends its metaID to the reader.
3. The reader looks for the corresponding key for the received metaID in its database and transmits the key the tag.
4. The tag hashes the received key and compares the result with the metaID. If the value is matched, the tag is unlocked and offers its full functionality to its reader in a short time period. Otherwise, no other functionality is offered.

The above hash-based access control prevents unauthorized readers from reading the tag contents due to the difficulty involved in inverting a one-way hash function [2]. Spoofing can be detected but it cannot be prevented, because an adversary can send a query to a tag asking for its metaID, and then the adversary can spoof the reader by sending this metaID when queried by the reader. The reader checks the consistency of the contents of the tag against the back-end database. When inconsistency is detected indicating that spoofing attacks have occurred, alerts can be generated.

Disadvantages of this scheme also include being capable of tracking individual tags because tags provide their unique nature to all the readers with their metaIDs, even in the locked state. Furthermore, we observe that this approach is vulnerable to the eavesdropping attack for the key.

3.5.1.2 Randomized Access Control

Randomized access control is proposed in Weis et al. [2] to extend the hash-based access control to overcome the predictable response of the tag by making the tag provide a randomized response as follows:

1. The reader sends a query to the tag.
2. The tag responds to the reader by sending $(r, hash(ID||r))$, where r is a random number.
3. The reader looks in its database for IDs using brute-force search and hashing each of them and concatenating with r until it finds a match with the value received from the tag.

The above approach is feasible only for a small number of tags due to the brute-force search. Furthermore, there is difficulty involved in inverting the function output, and it does not provide any guarantee about revealing the tag ID. To overcome this issue, under the assumptions that each tag shares a unique secret key k with the reader and a pseudo-random function ensemble $F = \{f_n\}_{n \in N}$ is supported, the tag replies to the query by sending $(r, ID \oplus f_k(r))$.

3.5.1.3 Silent Tree Walking

Because eavesdroppers can monitor downlinks with a longer distance, they can obtain a tag's ID by overhearing the reader's broadcasting each bit of the tag's ID when the tree-walking protocol is used. In the proposed silent tree-walking protocol [2], assume that there are tags that all share a common prefix, such as manufacturer ID. The reader asks tags to send their next bit. If all the tags share the same bit value, collision does not occur and the reader asks to send its next bit. If the collision takes place, the reader specifies which portion of the tags should proceed using the knowledge of the previous prefix. Because a long-range eavesdropper will not hear the tags' responses from the uplink, the eavesdropper does not know the prefix, but the reader can learn the prefix from the uplink. For example, there are two tags with ID values $a_1 a_2$ and $a_1 \bar{a}_2$. The reader receives a_1 from both the tags without collision, and detects the collision for the second bit because the two tags differ in the second bit. The reader uses the prefix knowledge a_1 (because it was received securely through the uplink) by sending either $a_1 \oplus a_2$ or $a_1 \oplus \bar{a}_2$ to specify which portion of the tags should proceed.

3.5.2 Authentication of Readers

To prevent hostile tracking and man-in-the-middle attacks, Gao et al. [3] exploit the property of randomized read access control to prevent reading from unauthorized readers by authenticating the readers before sending the tags' data.

This approach has the following assumptions:

1. Each reader has a unique ID denoted as RID.
2. Each tag includes two parts: the first part includes a ROM for storing its $hash(ID)$ and a RAM for storing the RID of an authenticated reader; and the second part is a logic circuit, performing computations such as the hash function.
3. The back-end database stores each tag's $(ID, hash(ID))$.

The approach has three steps. In the first step, the reader is authenticated; and in the second step, the reader obtains the tag ID from the tag's response and by referring to the database. The third step is about how to update the reader's ID. Authentication of a reader is conducted as follows:

1. The reader sends a query to a tag.
2. The tag generates a random number k and sends it to the reader.
3. The reader sends k to the back-end database.
4. The database sends the reader $hash(RID\|k)$.
5. The reader forwards $hash(RID\|k)$ to the tag.
6. The tag computes $hash(RID\|k)$ and compares it with the obtained one. If they match, the reader is authenticated successfully and the tag is ready to reveal its information related to its tag ID. Otherwise, the reader is not authenticated successfully. If an adversary obtains the value $hash(RID\|k)$, the adversary cannot use this value for further authentication, because the random number k will be changed the next time.

Obtaining a tag's ID is conducted as follows:

1. After successful authentication of the reader, the tag sends the $hash(ID)$ to the reader to prevent eavesdropping.
2. The reader looks for the pair $(ID, hash(ID))$ from the database and finds the tag's ID.

Take, for example, a supply chain application, when the objects are moved from one warehouse to another, the new reader's RID must be updated accordingly. Updating the reader's RID is achieved as follows:

1. The reader passes the tag's $hash(ID)$ obtained from the tag to the back-end database.
2. The database realizes that the RID stored in the tag must be updated.
3. The database finds the RID_{new} and transmits it to the reader.
4. The reader sends $RID_{new} \oplus RID_{old}$ to the tag.
5. The tag then obtains RID_{new} from XORing with RID_{old} to prevent spoofing, because an adversary cannot determine RID_{new} without knowing RID_{old}.

However, we found out that there are some problems in the above approach. First, updating the reader's RID cannot proceed because the new reader cannot be authenticated to obtain $hash(ID)$. Second, updating the reader's RID can be easily launched by an adversary to conduct DoS.

We did some revisions on their approach as follows:

1. The new reader tries to get authenticated by the tag, but failed.
2. The reader realizes that the RID stored in the tag must be updated. It informs the back-end database.
3. The database finds out the old RID, denoted as RID_{old}, using one of the following ways. An administrator can input manually or provide a list of candidate RIDs of the old reader. Otherwise, the back-end database maintains a list of candidate RIDs of the old reader. The database and the new reader work together to try to authenticate all possible RIDs of the old reader, one by one, until one old RID is authenticated by the tag.
4. The database transmits $RID_{new} \oplus RID_{old}$ to the reader, which passes $RID_{new} \oplus RID_{old}$ to the tag.
5. The tag then obtains RID_{new} by XORing with RID_{old} to prevent spoofing, because an adversary cannot determine RID_{new} without knowing RID_{old}.

The above approach is secure even when both readers and tags are eavesdropped.

3.5.3   Anti-Counterfeiting of RFID

The problem of counterfeiting is a problem encountered by several industries all over the world (e.g., copyright industry, clothing, fashion, automotive, medicines, etc.) and therefore leads to the loss of company revenues and various threats to public health and safety [4]. Some of the current anti-counterfeiting technologies used in companies include:

■ Optical anti-counterfeiting technologies such as holograms, retro-reflective material, and micro-printing technologies are some anti-counterfeiting technologies. However, manufacturing this equipment has become cheap and does not constitute a barrier for counterfeiters.
■ Microelectronics have received acceptance as anti-counterfeiting devices. However, they are expensive, and companies expect much cheaper anti-counterfeiting devices.

Furthermore, none of these technologies protect the products over a long period of time. Staake et al. [4] propose one solution: to use RFID along with the electronic product code (EPC) network. Using RFID technology in products can be traced from manufacturer to retailer. The unique product code is associated with a database entry. This track and trace in the EPC network provides information about goods that helps the consumer perform plausibility checks.

Assume that each tag has a key, a unique ID, and a cryptographic unit, and that the database stores the corresponding key in the EPC network [4]. The tag is authenticated as follows:

1. The tag sends its ID to the cryptographic unit (CU) at the manufacturer.
2. The CU sends a challenge message to the tag.
3. The tag sends the encrypted message (using its key) to the CU.
4. The CU looks up a database and verifies the response.

However, we observe that the above process is vulnerable to the eavesdropping.

3.5.4 Strengthening Tags against Cloning

Protecting low-cost RFID tags from various attacks is a challenging task due to the tags' low functionality. EPC tags do not support anti-cloning features [5]. EPC tags send their IDs to any reader that sends a query, and any reader accepts the ID — whatever it receives as a valid ID. Therefore, between the reader and tags, authentication is lacking, so that anyone with a reader can scan tags and can clone the tags with the obtained IDs.

Two types of tags are considered by Juels [5] to strengthen tags against cloning:

1. *Basic EPC tags* support only one security feature, called a privacy-enhancing kill command, which is PIN-protected (of 32 bits in length). When a tag receives a PIN, it sends "yes" or "no," indicating the validity of a kill PIN.
2. *Enhanced EPC tags* respond to a command called access that is accompanied by a valid 32-bit access PIN. When given this command, the tag enters a "secured state" and can allow access to the tag's memory only in this "secured" state.

Some mechanisms proposed by Juels [5] for anti-cloning are as follows:

1. *Authenticating basic EPC tags*: This simple scheme protects EPC tags against a skimming attack, which is defined as the process of scanning an EPC for the purpose of cloning an EPC tag. Let T_i, R, A, and K_i denote the unique index of an EPC tag: the reader, the tag, and the valid kill PIN for T_i, respectively. Let N denote the total number of tags. Let $PIN - test(K_i)$ denote the PIN-test with K_i,

and the output is b, where if b = "1" it means "valid," and otherwise it means "invalid." We have:

- $A : T \leftarrow T_i$
- $A \rightarrow R : T$
- R: if $T = T_x$ for some $1 \leq x \leq N$, then $i \leftarrow x$, else output "unknowing tag" and halt.
- $R \rightarrow A : PIN - test(K_i)$;
- $A \rightarrow R : b$
- R: if b = '1' then output "valid," Else output "invalid."

In the above approach, a trusted RFID device R attempts to authenticate tag A. A tag without a valid ID cannot be successfully authenticated. However, the above protocol has a basic vulnerability; that is, the tag can spoof the reader by simply accepting any PIN that is given by a reader, so that the protocol will always output "valid." To overcome this problem, the spurious PINs are used by the reader to test the responses of the tag as follows:

- $A : T \leftarrow T_i$
- $A \rightarrow R : T$
- R: if $T = T_x$ for some $1 \leq x \leq N$ then $i \leftarrow x$, else output "unknowing tag" and halt.
- R: $(j, \{P_i^{(1)}, P_i^{(2)} \ldots P_i^{(q)}\}) \leftarrow GeneratePINSet(i)[q]$; $M \leftarrow$ "valid" for $n = 1$ to q do
- $R \rightarrow A : PIN - test(P_i^{(n)})$
- $A \rightarrow R : b$
- R: if b = '1' and $n \neq j$ then $M \leftarrow$ "invalid"; if b = '0' and $n = j$ then $M \leftarrow$ "invalid"
- R: output M;

In the above scheme, q is the number of generated spurious PINs, GeneratePINSet generates a set of q-PINs, and one valid kill PIN K_i is inserted at a random position j, generated by the GeneratePINSet.

2. *Authenticating enhanced EPC tags:* For the enhanced EPC tags, the reader must transmit the access PIN to a tag in order to read its kill PIN. In the following approach, the access PIN is used to authenticate the reader, whereas the kill PIN is used to authenticate the tag, where B_i denotes the access PIN for tag i.

- $A : T \leftarrow T_i$
- $A \rightarrow R : T$
- R: if $T = T_x$ for some $1 \leq x \leq N$ then $i \leftarrow x$, $B \leftarrow B_i$, else output "unknown tag" and halt
- $R \rightarrow A : B$
- A: if $B = B_i$ then $K \leftarrow K_i$, else $K \leftarrow \phi$

- $A \rightarrow R : K$
- R: if $K = K_i$ then output "valid," else output "invalid."

An adversary can beat the above protocol by accepting the access PINs implicitly and by guessing the kill PIN, but the probability is small. However, we observe that this protocol is also vulnerable to the eavesdropping attack.

3. *Untrusted readers:* If the reader is not trustworthy, the following protocol can be used and it uses an entity called trusted entity V that has the knowledge of P_i.

- $A : T \leftarrow T_i$
- $A \rightarrow R \rightarrow V : T$
- V: if $T = T_x$ for some $1 \leq x \leq N$ then $i \leftarrow x$, else output "unknown tag" and halt
- $V: (j, \{P_i^{(1)}, P_i^{(2)} \ldots P_i^{(q)}\}) \leftarrow GeneratePINSet(i)[q])$
- $V \rightarrow R : \{P_i^{(n)}\}_{n=1}^{q}$

For $n = 1$ to q do
- $R \rightarrow A : PIN - test(P_i^{(n)})$
- $A \rightarrow R : R^{(n)}$

- $R \rightarrow V : \{R^{(n)}\}_{n=1}^{q}$
- $V: M \leftarrow$ "valid";
  For $n = 1$ to q do
  {if $R^{(n)} = $ '1' and $n \neq j$ then $M \leftarrow$ "invalid"
  if $R^{(n)} = $ '0' and $n = j$ then $M \leftarrow$ "invalid"}
  Output M

In the above scheme, if an attacker has compromised the reader R and knows T_i, the attacker cannot determine whether the presented PIN is correct or not without accessing the tag i. Therefore, the attacker needs to guess the correct PIN from a set of q PINs with the probability $1/q$.

3.5.5 The Blocker Tag

One approach to improve privacy is the "Kill Tag" approach, in which tags are killed before they are placed in the hands of consumers. Killing tags is a PIN-based operation that protects against unwanted access. However, the killed tags cannot be reactivated [6]. Another approach is called the Faraday cage approach, in which RFID tags are shielded from scrutiny using a container made of metal mesh or foil called a Faraday cage that is impenetrable by radio signals. However, not all items can be placed in containers (e.g., human beings, pets, etc.).

In Juels [6], a concept called "blocker tag" is proposed. In the tree-walking protocol, the reader queries the tags in the sub-tree of a given node B for their next bit value, and the tags send their next bit value. The blocker tag approach uses the tree-walking protocol so that a blocker tag broadcasts both the '0' and '1' bits simultaneously instead of sending just one bit at a time, so that a reader get confused and cannot determine which ID belongs to which tag. To accomplish this, blocker tags require two antennae. This collision makes the reader recurse on all nodes by making the reader to explore the entire tree. In this way, the full-blocker tag blocks the reading of all tags due to time constraints.

However, blocker tags can be also used as attacks. A refined blocker tag blocks only a certain subset of tags and the tag is called a "partial blocker" tag or a "selective blocker" tag. Instead of using the blocker tag concept to block all the tags, the selective blocker concept can be used to block only the tags that lie within a certain range. It can be used to obstruct only the reading of tags that begin with '0' bit in its serial number and the tags whose ID starts with '1' can be read without any interference.

3.6 Conclusions

This chapter surveyed some security and privacy issues that a low-cost RFID tag encounters and discussed some security and privacy solutions proposed in the literature. The challenge is that security is difficult to achieve because tags are very low cost. We also point out some problems in some of these solutions and propose some enhancements. Further research is needed in this area.

References

1. Y. Xiao, S. Yu, K. Wu, Q. Ni, C. Janecek, and J. Nordstad, "Radio frequency identification: technologies, applications, and research issues," *Journal of Wireless Communications and Mobile Computing (JWCMC)*, 2006, accepted for publication.
2. S. Weis, S. Sarma, R. Rivest, and D. Engels, "Security and privacy aspects of low-cost radio frequency identification systems," *Proc. of International Conference on Security in Pervasive Computing*, March 2003.
3. X. Gao, Z. Xiang, H. Wang, J. Shen, J. Huang, and S. Song, "An approach to security and privacy of RFID system for supply chain," *Proc. of Conference on E-Commerce Technology for Dynamic E-Business,* September 2005.
4. T. Staake, F. Thiesse, and E. Fleisch, "Extending the EPC network — The potential of RFID in anti-counterfeiting," *Proc. of IEEE Symposium on Applied Computing — SAC,* March 2005.
5. A. Juels, "Strengthening EPC tags against cloning," manuscript, March 2005.

6. A. Juels, R. Rivest, and M. Szydlo, "The blocker tag: selective blocking of RFID tags for consumer privacy," *Proc. of Conference on Computer and Communications Security — ACM CCS*, October 2003.
7. A. Juels and J. Brainard, Soft blocking: flexible blocker tags on the cheap, Workshop on Privacy in the Electronic Society — WPES, October 2004.
8. A. Juels and R. Pappu, "Squealing euros: privacy protection in RFID-enabled banknotes," *Proc. of Financial Cryptography,* January 2003.
9. G. Avoine, "Privacy issues in RFID banknote protection schemes," *Proc. of International Conference on Smart Card Research and Advanced Applications — Cardis,* August 2004.
10. A. Juels, "Minimalist cryptography for low-cost RFID tags," *Proc. of International Conference on Security in Communication Networks — SCN,* September 2004.
11. S. Weis, "Security and Privacy in Radio-Frequency Identification Devices," Master thesis, May 2003.
12. M. Ohkubo, K. Suzuki, and S. Kinoshita, "Cryptographic approach to 'privacy-friendly' tags," *Proc. RFID Privacy Workshop,* November 2003.

Chapter 4

Watermarking Technique for Sensor Networks: Foundations and Applications

Farinaz Koushanfar and Miodrag Potkonjak

Contents

4.1   Abstract

We propose the first watermarking approach for protecting data and information generated in wireless embedded sensor networks. Contrary to current state-of-the-art image and audio watermarking, we embed signatures before or during the process of data acquisition and processing. The key idea is to impose additional conditions that correspond to the cryptographically encoded author/owner's signature during the data acquisition or processing in such a way that the quality of recording or processing is not or is only minimally impacted.

The technique is generic in that it is applicable to a number of tasks in WASNs, including sensor design, deployment, and data fusion. It enables real-time watermark checking and can be used in conjunction with a variety of watermarking protocols. It can also be applied to any collected multi-modal data or data generated for simulation purposes. Finally, the technique is adaptive, optimization-intensive, and resilient against attack. Through simulation and prototyping we demonstrate the effectiveness of active watermarking on two important sensor network tasks: (1) location discovery and (2) light characterization.

4.2   Introduction

Wireless ad hoc (multi-hop) networks are a new, emerging paradigm for bandwidth and energy-efficient wireless systems. In multi-hop networks, each wireless node communicates only with a few adjacent wireless nodes to enable a low-power communication. The network is ad-hoc in that each node operates autonomously and in an unattended manner. The addition of sensors to a wireless network enables the system to integrate the physical world into computations. The integration is made possible by sensing and measuring different dimensions of a physical phenomenon, storing the measurements as data, and then collaboratively processing the stored data. To capture a physical phenomenon in a sound way, the network should contain a vast amount of information and data flowing among the wireless nodes.

A large number of sensor network applications are security critical and therefore processing and flow of the data in the network must be recorded. For example, sensor networks can provide comprehensive information about all people and their activities in an instrumented environment. Typical security-critical applications include intelligent security systems in smart buildings, detection and tracking in the battlefield, and secure pervasive computing. To address these challenges, we have developed, to the best of our knowledge, the first technique for data authentication in sensor networks. The idea is to add invisible marks to data and information in such a way that the accuracy of data is only nominally impacted. The marks provide the proof that particular data and information is indeed produced by a particular node.

Although a number of conventional cryptography protocols already exist for wireless communication networks, the inherent constraints in sensor networks (e.g., energy constraints, small size, limited computation), along with the enormous size of the data, make the use of the traditional techniques impractical. To address these issues, a suite of security protocols for sensor networks called SPIN was recently proposed [22]. The SPIN family of protocols permits only valid key holders access to encrypted data; but as soon as this data is decrypted, tracking the reproduction or re-transmission of the data is not possible. Digital watermarking of the processed data complements the security need addressed by security protocols such as SPIN. A watermarking scheme embeds a digital signature into the data, where the data can be the outcome of a process, images, audio, text, or video signals.

We introduce an active watermarking technique that can be applied on the data that is processed during a sensor fusion application. These types of applications are typically known as multimodal sensor fusion (MMSF) [5]. In MMSF, data from different types of sensors and different dimensions is gathered (fused) and then collaboratively processed to capture different aspects of a physical reality. To have a meaningful interpretation of the fused data, the system should have a model of physical reality into which the data is mapped. A more detailed discussion of MMSF and its different models are presented in Section 4.6.3.

Probably the best way to introduce the multimodal sensor fusion (MMSF) watermarking approach is to use a small, yet illustrative example. For this purpose, we use *atomic acoustic trilateration*. Atomic acoustic trilateration is used for locating an unknown wireless node, by measuring its distance to a few already located nodes (beacons) and applying trilateration models. The distances are estimated by measuring the travel time of an acoustic signal between the nodes. The atomic acoustic trilateration method has been widely accepted and used in sensor network location discovery systems [12,23]. For the sake of conceptual simplicity, we show a two-dimensional (2D) case. The generalization to the 3D case that is used in our experimental results is straightforward.

Trilateration is the procedure for determining the spatial coordinates of a node, given the erroneous measured distances of the node to beacons. A typical scenario is shown in Figure 4.1. Node X trilaterates with three beacon nodes A, B, and C that have coordinates (x_A, y_A), (x_B, y_B), and

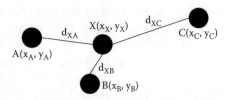

Figure 4.1 Atomic trilateration procedure. Node X estimates its spatial coordinates using beacon nodes A, B, and C.

(x_C, y_C), respectively. The distance is estimated using time differences of arrival (TDoA) between radio frequency (RF) and acoustic signals simultaneously emitted from a beacon node and received at the node X [11]. Node X turns on a timer upon receiving the RF signal from a beacon to measure the difference between the arrival of the RF and acoustic signals from that beacon. The timer measurements have an intrinsic error. The speed of the acoustic signal is a function of the temperature of the propagation media. Equation (4.1) states the relationship between the speed of the acoustic signal V_s (m/s) and the temperature T_c (°C):

$$V_s = 331.4 + 0.6T_c \tag{4.1}$$

If we assume the exact timer measurements Δt_{XA}, Δt_{XB}, and Δt_{XC} between the node X and the beacon nodes, the distances d_{XA}, d_{XB}, and d_{XC} between node X and the beacon nodes are then measured using TDoA Equations (4.2):

$$d_{XA} = V_s.\Delta t_{XA}$$

$$d_{XB} = V_s.\Delta t_{XB} \tag{4.2}$$

$$d_{XC} = V_s.\Delta t_{XC}$$

The coordinates of $X(x_X, y_X)$ are then computed using Euclidean Equations (4.3):

$$d_{XA} = \sqrt{(x_X - x_A)^2 + (y_X - y_A)^2}$$

$$d_{XB} = \sqrt{(x_X - x_B)^2 + (y_X - y_B)^2} \tag{4.3}$$

$$d_{XC} = \sqrt{(x_X - x_C)^2 + (y_X - y_C)^2}$$

In an error-free world, computing V_s from Equation (4.1) and then replacing the value of the distances from Equations (4.2) into Equations (4.3), would yield the coordinates (x_X, y_X). However, there are intrinsic inaccuracies in measuring T_c in Equation (4.1) and in measuring Δt_{XA}, Δt_{XB}, and Δt_{XC} in Equation (4.2). These inaccuracies change the nature of the problem from a nonlinear system of equations with a unique solution to an optimization problem, where the goal is to minimize the errors in the system of equations. Next, we show the formulation of Equations (4.1), (4.2), and (4.3) as a nonlinear optimization problem in terms of $(\epsilon_T, \epsilon_{XA}, \epsilon_{XB}, \epsilon_{XC}, \delta_1, \delta_2, \delta_3)$. The terms $\epsilon_T, \epsilon_{XA}, \epsilon_{XB}, \epsilon_{XC}$ denote the errors in the measurement of $T_c, \Delta t_{XA}, \Delta t_{XB}$, and Δt_{XC}, respectively. Also, the variables $\delta_1, \delta_2, \delta_3$ are the errors between the Euclidean distances measured in Equations 4.3 and the TDoA measured distances of d_{XA}, d_{XB}, d_{XC} in Equations 4.2, respectively. The objective function is to minimize the overall error in the system, and can be stated as shown in Equation (4.4):

$$\min\ F(\epsilon_T, \epsilon_{XA}, \epsilon_{XB}, \epsilon_{XC}, \delta_1, \delta_2, \delta_3) \tag{4.4}$$

Constraints are necessary for bounding the discrepancies between Equations 4.2 and 4.3. The constraints are labeled c_1, c_2, and c_3, and are shown in Equations (4.5):

$$c_1 : |\sqrt{(x_X - x_A)^2 + (y_X - y_A)^2}$$
$$- (331.4 + 0.6(T_c + \epsilon_T))(\Delta t_{XA} + \epsilon_{XA})| \leq \delta_1$$

$$c_2 : |\sqrt{(x_X - x_B)^2 + (y_X - y_B)^2}$$
$$- (331.4 + 0.6(T_c + \epsilon_T))(\Delta t_{XB} + \epsilon_{XB})| \leq \delta_2$$

$$c_3 : |\sqrt{(x_X - x_C)^2 + (y_X - y_C)^2}$$
$$- (331.4 + 0.6(T_c + \epsilon_T))(\Delta t_{XC} + \epsilon_{XC})| \leq \delta_3 \tag{4.5}$$

We propose a watermarking technique that embeds the signature of the processing node (i.e., the node that performs an optimization computation) into the outcome of the optimization. The outcome is sent to the other nodes in the network for many purposes, and also could be used to identify its computing node (the author). The key idea of our watermarking technique is to embed a signature into solutions of a nonlinear optimization problem by preprocessing the formulation of the problem before solving it. For example, we can alter the objective function or add additional constraints that correspond to the author's signature watermark. In this particular example, we show three conceptually different methods for embedding the following stream of 24 bits into the solution:

$$110100000111011010010010$$

These methods are: (1) addition of new constraints, (2) augmentation to the objective function, and (3) addition of new variables. Although we illustrate only one instance for each method, the methods are very general and contain different variations for embedding the signature based on the preprocessing procedure of choice. Also, one can combine these methods for embedding a particular signature into the optimization procedure.

4.2.1 Addition of New Constraints

To add watermarking constraints, we use a linear combination of the variables in the optimization objective function. We number the variables using a particular low of the Kolmogorov complexity rule as shown in Table 4.1. Next, we group the watermark stream into groups of seven bits (in this case, seven is the number of variables):

$$\overbrace{1101000}, \overbrace{0011101}, \overbrace{1010010}, \overbrace{010}$$

We further number the bits within each group from 1 to 7 from left to right (in this case, the last group will have only three members). We then match the bit numbers with the corresponding variable numbers from Table 4.1. For each group of seven or less variables, we add a new constraint to the already existing constraints of the problem. If a bit zero is assigned to a variable within a group, that variable is not included in the linear combination of the variables from that group. According to these rules, we add the following four constraints c_4, c_5, c_6, and c_7 to the above.

$$c_4 : \epsilon_T + \epsilon_{XA} + \epsilon_{XC} \leq v_1$$

$$c_5 : \epsilon_{XB} + \epsilon_{XC} + \delta_1 + \delta_3 \leq v_2$$

$$c_6 : \epsilon_T + \epsilon_{XB} + \delta_2 \leq v_3$$

$$c_7 : \epsilon_{XA} \leq v_4 \tag{4.6}$$

The values of v_1, v_2, v_3, and v_4 are selected such that the feasibility of the solution space of the optimization problem is not harmed. They also must

Table 4.1 Assignment of Numbers to Variables of the Optimization Problem

1	2	3	4	5	6	7
ϵ_T	ϵ_{XA}	ϵ_{XB}	ϵ_{XC}	δ_1	δ_2	δ_3

guarantee that the probability of a coincidental solution that satisfies the same watermark is low.

4.2.2 Augmentation to the Objective Function

The original optimization problem consists of an objective function (Equation 4.4) and a set of constraints (Equations 4.5). Another way to encode the signature within the optimization problem, and thus to put a watermark in the signature, is to augment to the objective function. The initial goal of the objective function is to minimize the error in the system of equations, but how the overall error is defined may differ. Our particular way of embedding the message into the objective function is by adding terms corresponding to the codeword, using the enumeration of variables shown in Table 4.1. Specifically, because we have seven variables, again we use the separation of the signature into groups of seven. For each group, we add a set of terms to the objective function. For example, for the selected codeword, we change the OF by adding terms as shown in Equation (4.7). In this example, whenever bit i equals one (within a seven bit group), we add the difference between the corresponding variable i and the next variable $(i + 1)$. The advantage of adding the difference is that ultimately minimizing the objective function would result not only in minimizing a function of errors (F), but also it would result in balancing the value of different types of errors.

$$\min F(\epsilon_T, \epsilon_{XA}, \epsilon_{XB}, \epsilon_{XC}, \delta_1, \delta_2, \delta_3)$$

$$+ \quad F_1(|\epsilon_T - \epsilon_{XA}| + |\epsilon_{XA} - \epsilon_{XB}| + |\epsilon_{XC} - \delta_1|)$$

$$+ \quad F_1(|\epsilon_{XB} - \epsilon_{XC}| + |\epsilon_{XC} - \delta_1| + |\delta_1 - \delta_2| + |\delta_3 - \epsilon_T|)$$

$$+ \quad F_1(|\epsilon_T - \epsilon_{XA}| + |\epsilon_{XB} - \epsilon_{XC}| + |\delta_2 - \delta_3|) + F_1(\epsilon_{XA} - \epsilon_{XB}|) \quad (4.7)$$

For the differences to be as small as possible and also to assist in the verification of the watermarks, we add new constraints corresponding to the additional parts in the objective function. The corresponding constraints are labeled $\hat{c}_4$, $\hat{c}_5$, $\hat{c}_6$, and $\hat{c}_7$, as shown in Equations 4.8, and are added to the original constraints illustrated in Equations 4.5.

$$\hat{c}_4 : (|\epsilon_T - \epsilon_{XA}| + |\epsilon_{XA} - \epsilon_{XB}| + |\epsilon_{XC} - \delta_1|) \leq \hat{v}_1$$

$$\hat{c}_5 : (|\epsilon_{XB} - \epsilon_{XC}| + |\epsilon_{XC} - \delta_1| + |\delta_1 - \delta_2| + |\delta_3 - \epsilon_T|) \leq \hat{v}_2$$

$$\hat{c}_6 : (|\epsilon_T - \epsilon_{XA}| + |\epsilon_{XB} - \epsilon_{XC}| + |\delta_2 - \delta_3|) \leq \hat{v}_3$$

$$\hat{c}_7 : (\epsilon_{XA} - \epsilon_{XB}|) \leq \hat{v}_4 \quad (4.8)$$

Again, the relationships $\hat{c}_4$, $\hat{c}_5$, $\hat{c}_6$, and $\hat{c}_7$ are used along with the values of $\hat{v}_1$, $\hat{v}_2$, $\hat{v}_3$, and $\hat{v}_4$ by the authorizing parties to verify the existence of a watermark.

4.2.3 Addition of New Variables

To enable the watermarking procedure to encode larger messages without over-constraining the original problem, we can add auxiliary variables to the problem. In the example in Figure 4.1, we add five auxiliary variables A_1, A_2, A_3, A_4, and A_5. For example, we can have a new partition of the variables into groups of 7 and 5, as shown below:

$$\overbrace{1101000}, \overbrace{00111}, \overbrace{0110100}, \overbrace{10010}$$

The new corresponding objective function and the additional constraints c_1', c_2', c_3', and c_4', are shown in Equations (4.9) and (4.10), respectively.

$$\min F(\epsilon_T, \epsilon_{XA}, \epsilon_{XB}, \epsilon_{XC}, \delta_1, \delta_2, \delta_3) + F'(A_1, A_2, A_3, A_4, A_5). \quad (4.9)$$

$$c_4' : \epsilon_T + \epsilon_{XA} + \epsilon_{XC} \leq v_1',$$

$$c_5' : A_3 + A_4 + A_5 \leq v_2',$$

$$c_6' : \epsilon_{XA} + \epsilon_{XB} + \delta_1 \leq v_3',$$

$$c_7' : A_1 + A_4 \leq v_4'. \quad (4.10)$$

Thus far, we have shown different preprocessing methods for embedding a watermark into solutions of a typical optimization problem used for MMSF. We conclude this section by restating the challenges of watermarking. Watermarking challenges can be broadly classified into two groups: (1) those that are common for all high-quality watermarking techniques and (2) those specific to watermarking in sensor networks. The main generic watermarking challenges can be traced along four lines: (1) quality of watermarked solution, (2) quality of watermarking process, (3) interaction with standard tools, and (4) complexity of practical implementation/realization. Quality of solution is measured by the accuracy of a watermarked versus the non-watermarked solution. Quality of watermark is the strength of the authorship proof and is measured by coincidence probability, resiliency against ghost signatures, and watermarking attacks. Coincidence probability is the probability that one would generate a watermarked solution although no watermarking is really conducted. U.S. courts accept probabilities as large as 10^{-6} and definitely 10^{-9} as clear indication that coincidence does not pose reasonable doubts. The third group of criteria mainly refers to the need for transparency or the minimal need for changing standard design or operation procedure and tools. Finally, the procedure must be simple enough in order not to impose significant additional development cost. Other performance measures for a watermark include runtime, low additional required storage, and, in the case of sensor networks, low communication overhead.

We have developed a family of watermarking techniques that can be applied during the design of sensor network elements, their deployment, data acquisition, and data processing. The key common denominator of all these watermarking schemes is that we always impose additional conditions on how a particular design or operation step is conducted. The conditions correspond to the cryptographically encoded signature of the author/owner. The key premise is that there are numerous ways to design and deploy sensor network nodes and record and process data that results in producing data of very similar utility.

The major features of our watermarking approach are that the techniques are active, generic, quantitative (optimization intensive), transparent, fair, flexible, and real-time. The introduced watermarking technique is *active* because it is conducted during data acquisition/processing and therefore leverages the key advantage that the author/owner has over attackers. Traditional multimedia watermarking is done as a postprocessing step when data is already recorded and where this key authors' advantage is lost. The technique is *generic* in that it can be applied at all levels of design and operation of the sensor network, regardless of type of sensor used and the method of data processing. The technique is *quantitative* in that for given data and a given signature, the watermarking procedure conducts a search for the solution that satisfies both criteria. The scheme is *flexible* because it is easy to explore the trade-off between the quality of the solution versus the quality of watermarking. Because the technique is done as preprocessing, it is *fully transparent* with multimodal sensor data fusion. Finally, it has very minimal runtime and communication overhead. In particular, checking for a watermark on a given data set can be done very rapidly and without needing access to the original data.

4.3   Preliminaries

We briefly survey the key abstractions required for the systematic and sound design and deployment of sensor networks and therefore also for watermarking or any other IPP technique. We also outline specific sets of models used by our two demonstration studies: atomic trilateration and light source characterization.

We have developed models for the physical world, phenomenon, sensors, and computation. The physical world abstractions are mainly related to concepts of space and time and their interaction. The main premise regarding space is that we follow a 3D Euclidean space model.

The space abstraction's components include the maximal resolution of distance measurements and the size of the space instrumented by a sensor network. The most important aspect of time in the case of distributed sensor networks is the assumed mechanism of synchronization of multiple entities. We assume that synchronization between the nodes in the network is achieved with a certain error distribution.

Phenomenon measured by sensors is characterized using a set of equations. For atomic trilateration, we use the relationship between the speed of sound and temperature. We assume that the radio frequency (RF) signal immediately arrives at the receiver and serves as a time synchronizer between the sender and the receiver. This is a reasonable assumption, because the precision of all measurements is limited. For light, we assume that each source is a point source and has four components required for its characterization: intensity and three location coordinates.

For all sensors, we assume 10-bit resolution and specific error models as stated in Experimental Results (Section 4.8). For the computational model, we follow fully periodic synchronous data flow (SDF). We assume that all computations are done using the IEEE floating point standard. In our experimentation, we consider data from one to five iterations of the SDF model.

4.4   Related Work

Both sensor networks and watermarking for intellectual property protection are widely studied research topics with broad agendas and a great variety of approaches. We briefly survey only the most relevant efforts in sensor networks and intellectual property protection.

Sensor networks (SNs) have emerged as one of the most attractive and powerful directions in ubiquitous computing. They provide a versatile and economically viable interface between computer systems and the physical world. SNs pose a number of demanding new technical problems, including low power designs [26], operating systems [13], and localized algorithms.

Due to the exponential proliferation of the Internet and technologies and software for inexpensive and rapid copying of data and software, intellectual property protection (IPP) has emerged in the last decade as an important area of research and development. Recently, a variety of IPP techniques have been proposed, such as cryptography [9], fingerprinting [1], obfuscation [7], copy detection [4], and reverse-engineering [3].

Watermarking is a process of embedding hidden data into an object in order to facilitate identification of the creator or owner in such a way that the perception or functionality of the object is preserved [8,15,27]. Watermarking techniques can be broadly classified into three groups.

The first group consists of post-processing techniques that embed signatures into recorded data artifacts [29] and that have only a syntactic component that is not altered during its use. Examples include water-marking of images [28], video [19], and audio [16]. These techniques use post-processing methods to embed a particular message into the object by leveraging the imperfection of human perceptory systems. The second group of techniques embeds watermarks in computer-generated data in such a way that their visual appearance is not altered. Techniques have been proposed for the watermarking of textual objects [2], and computer generated static models [24] and dynamic (animated) [6] artifacts. The third group consists of pre-processing watermarking techniques for functional artifacts such as integrated circuit (IC) designs and software at different levels of abstraction. For example, some of the targeted levels for ICs include system and behavioral synthesis, physical design, and logic synthesis. These techniques mainly leverage the fact that there are often numerous solutions of similar quality for a given optimization and that it is possible to generate one that has certain characteristics corresponding to the designer's signature. More complex watermarking protocols, such as multiple watermarks [20], fragile watermarks [10], and publicly detectable watermarks [25], have also been developed.

Watermarking of sensor data follows unique requirements and concep-tually new information-hiding principles and techniques. It is distinctive from watermarking multimedia data such as audio, image, and video be-cause the final users of the sensor data are applications that have specified levels of acceptable distortion. Furthermore, the data is subject to process-ing by MMSF nonlinear equation solvers. Also, the amount of data can vary from simple scalar temperature measurements to complex vector variables of light sensors to exceptionally complex multimodal systems. For exam-ple, data rates can vary from less than one sample per hertz to more than a billion samples per second. It is also different from watermarking soft-ware and hardware because it is associated with continuous problems and input data is noisy. To address these requirements, we have developed the first generic concept of how to embed watermarks into the solution of a nonlinear system of equations.

4.5 Watermarking for Sensor Networks: Technical Concepts

There are a number of ways and places for embedding a watermark in an embedded wireless ad hoc sensor network. For example, a mark can be placed as a permanent part of a sensor or can be embedded in a data processing step during network operation. Our primary research goal is to introduce a system of techniques for embedding watermarks in data and in-formation produced by an ad hoc embedded sensor network. However, the

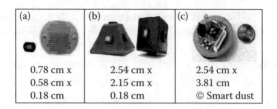

(a)	(b)	(c)
0.78 cm x 0.58 cm x 0.18 cm	2.54 cm x 2.15 cm x 0.18 cm	2.54 cm x 3.81 cm © Smart dust

Figure 4.2 (a) A single sensor, (b) a packed sensor, and (c) a wireless node equipped with packed sensors. The corresponding sizes are also shown.

exceptionally broad scope of the subject does not allow us to go through the details of each watermarking method, at each level of the sensor network. Therefore, we limit the scope of the technical part of this chapter to watermarking of data during the data processing step. In this section, however, we briefly introduce techniques that answer the questions of where and how to put a watermark in a sensor network. Figure 4.2 shows several components (light sensors, system of lights sensors, and node) of a sensor network. We start our presentation with the basic building block: the sensor. Next, we introduce watermarking techniques for the raw data acquired from the environment. Techniques for watermarking final information provided by the applications as the result of the data processing phase are presented in the next section. The most basic part of a sensor network is the sensor itself. The sensor provides a means for the network to interact with the physical environment to gather raw data. Even when we consider only a single sensor, we can add a watermark in many different ways. For example, we can add a watermark to data recorded by a single sensor in one of the following ways:

- *Manufacturer's watermark:* The manufacturer of the sensor can build a sensor in a unique way that could affect all the data acquired by that particular sensor. For example, in the case of the light sensor (miniature silicon solar cell) shown in Figure 4.2a, some of the photoelectric pixels might be removed or intentionally damaged in order to place a watermark on the light measurements from that sensor.
- *Calibration:* Once a sensor is manufactured, it must be calibrated to normalize the measurements of all sensors and to ensure proper functionality. Calibration is a complex process that can be formulated as optimization of an objective function and constraints. Therefore, we can treat the calibration in the same way as we treated acoustic atomic trilateration to embed watermarks.
- *Deployment:* The deployment method and the spatial coordinates and orientation of the sensor can greatly impact the data acquired

by that sensor. For example, in light sensors, the direction of the sensor has a direct impact on the intensity of the sensed light in a particular direction.

■ *Operation:* The details of operation of one sensor can also serve as a watermark on the individual or sequence of data gathered by that sensor. For example, the specific sampling moments of data or the specific regularity in sampling rate can be used as a mark. Also, adding small jitter on time sampling moments data can have a signature effect on the data. A major concern with the watermark is how much it affects the quality of the results. Note that all the stated techniques actually have very minimal impact on the final results. All that watermarking impact is a particular way how the experiment (measurements) are realized and selects one that has equally good chances as other to extract the correct final information.

After a set of sensors acquires the raw data from the environment, the wireless node manages the data within the nodes. This data management is application dependent and varies from one type of sensor network to the other. For example, if a node has multiple sensors, the ordering of sampling events of different sensors can be used as a mark. Also, for some applications, a number of wireless nodes need to communicate to exchange data; a mark can be embedded into the grouping and communication schemes of different nodes. We briefly summarize several conceptual ideas for watermarking at this level:

■ *Node coordinations and synchronization:* Not only the spatial coordination of a sensor, but also the spatial coordination of a node will affect the data acquired by the sensors on that node. Also, relative coordination of the sensors would affect the data gathered from groups of sensors. Synchronization is also very important because most often, nodes are tracking real-time phenomenon and are measuring different aspects of an event. The way in which synchronization is conducted can be used as a mark for different operations in the sensor network.

■ *Grouping and data transfer between the nodes:* Collaboration between the nodes and data transfer among them is a crucial task in sensor networks that can be used for many applications, such as routing, location discovery, and standby strategy. The collaboration scheme between groups of sensors can be used as a signature for the network. For example, defining that subset nodes cannot be in standby mode simultaneously, or a subset of nodes that can communicate but cannot be on the same route, can be used as a watermark for the operation of a sensor network.

■ *Data storage and processing:* A sensor network is a system of numerous nodes, where each node has a number of the following capabilities: sensing, storage, communication, computation and maybe actuation. For a particular task, not all the participating nodes need to compute and process the data, and not all of them need to store the data. The location of data storage or processing can be used as a mark for a data management scheme.

4.6 Generic Flow

In this section we introduce the active watermarking scheme. Figure 4.3 shows the generic flow of the watermarking process. The top right-hand side of the figure consists of two components and indicates the original flow of the sensor data. Specifically, the middle flow shows how sensors acquire data from the environment and then process that data into a format suitable for multi-modal sensor fusion. The other flow, on the right-hand side, indicates development of the model (objective function and constraints) that captures the relationship between sensor data and the value of data that is of the interest to the user. Note that the model is usually developed and entered only once, either before or after the deployment of the embedded sensor network. The top left-hand side of the figure shows a pre-processing method to prepare a watermark and then to convert it to the MMSF model format. The technical details of this part of the flow are shown in much

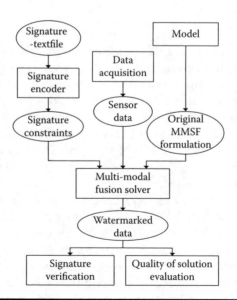

Figure 4.3   Generic flow of the watermarking process for a multi-modal sensor fusion (MMSF) task.

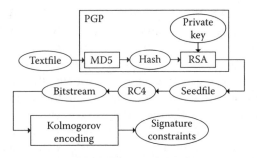

Figure 4.4   Signature encoding — basic building block.

more detail in Figure 4.4. The input is the signature of the owner. The signature is encoded using the standard cryptographic technique in order to ensure resiliency against statistical and pattern recognition attacks and is consequently translated into an MMSF model format. Specifically, each block of a relatively small number of consecutive random bits corresponds to particular watermarking constraints. The next step is integration of the original MMSF format and the pre-processed watermark, and their entry into a MMSF solver. The output of the MMSF solver is the watermarked information. Figure 4.3 also shows the process by which an authorized verifier can verify the presence of a signature and the way to verify the quality of the watermarked solution. The details of the verification process are shown in Figure 4.5.

4.6.1   Signature Preparation, Validation, and Detection

Signature preparation is the process that interprets the user's signature from a human perceptible format (usually a text message) into the format that is suitable for integration within the physical world model. The signature is encrypted so that the constraints do not have any statistical regularity. This phase is done before the signature is encoded into the specific format of the problem to prevent attacks such as signature detection or signature forging.

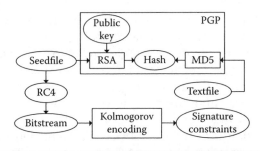

Figure 4.5   Signature verification — basic building block.

Figure 4.4 illustrates the encoding process that we adopted. We use the PGP software package shown in the large rectangle at the top of the figure. In the PGP package, MD5 is a one-way hash function and RSA is a popular public key encryption system. We use a stream cypher RC4 to produce a cryptographically strong pseudo-random bitstream. Finally, the low-complexity Kolmogorov encoding unit interprets the code in the proper objective function or constraints formats of the model.

Figure 4.5 illustrates the verification process to check for the presence of the watermark. For verification, we must show that the watermark corresponds to the original text file on the left-hand side of Figure 4.4. We must also confirm the presence of the signature in the solution of the MMSF problem. In the case of the objective function-based and additional variable-based watermarking schemes, the presence of the signature is proven by showing that it satisfies enough constraints such that the probability of coincidence is extremely low. In the case of constraint-based watermarking schemes, the presence of a signature is proven by satisfaction of all the constraints. We again use the PGP-based flow to show if the signature corresponds to the original text file and to the author's public key.

Note that there are no limitations on which RSA keys or text messages should be used, thus yielding a large selection of possible protocols. Otherwise, it might be possible to find false signatures using a brute-force attack. Also, like any other watermarking protocol, the verification procedure should have a probabilistic proof that the likelihood of other data having the same watermark is extremely low (e.g., less than 10^{-10}).

4.6.2   Resiliency against Watermarking Attacks

One of the primary criteria for evaluation of any cryptographic, security, and intellectual property protection technique is its resiliency against potential attacks. In this subsection we discuss the four most common types of attacks on watermarks (ghost signatures, addition of a new signature, deletion of the initial signature, and de-synchronization).

4.6.2.1   Ghost Signatures

The goal of this attack is to announce the presence of a signature in the watermarked solution that was not intentionally placed. There are two ways the attacker can try to identify and announce the existence of his signature. The first one is starting from the solution characteristics and trying to figure out the input that must be applied for generating the current solution. The potential for success of this approach is low, because it requires traversing through a one-way function in the inverse direction.

The other option is to try a number of different signatures, hoping to find a signature that will map into the existing solution. Because the probability of collision of two solutions after going through the message digest technique is exceptionally low, there is no realistic chance to successfully conduct this type of attack. For additional protection, we suggest taking the binary representation of measured values of sensor data and conducting an XOR ($\otimes$) function with the binary representation of RC4 output before the generation of watermarking constraints. In such a way, the author will place additional burden on the attacker in terms of the number of computationally intractable tasks that must be solved for a successful attack. Note that this augmentation imposes a need for synchronization during the signature verification process.

4.6.2.2 Addition of a New Signature

This type of attack attempts to add a new signature by altering the existing watermarked solution. While the addition of a new signature cannot be prevented, this attack is not effective because the author can easily show the solution (data) that has only his watermark, while the attacker does not have that type of solution. As an additional defense mechanism, the author can release only a subset of data or create the watermarked solution that has a large percentage of values on the boundary of maximally tolerable infidelity. In the first case, the attacker would not be able to generate the required consistent data. An analogy can better illuminate the idea. If an author releases only two thirds of each photograph, the attacker will have serious difficulty generating missing parts of a large set of such photographs economically. Note that often two thirds or so of a photograph can have similar economical value for many users as the whole picture. The latter augmentation mechanism prevents the alternation of the solution (data) because the attacker risks that any alternation can result in a non-usable data format.

4.6.2.3 Removal of the Author's Signature

The idea of an attacker might be to alter data sufficiently so that the integrity of the signature of the author and therefore its strength of proof are sufficiently reduced. Because the signature after the signature encoding process is fully randomized, the attacker cannot use statistical methods to identify the components of the signature. Because the signature has many independent components (constraints), even alteration of the solution to the extent that a significant percentage of them is violated is not sufficient. For example, if the signature consists of 400 constraints, even only 100 non-altered would provide strong proof of authorship because the likelihood that they are satisfied in a non-watermarked solution is extremely low. As additional resiliency mechanisms, we propose the use of error correction code to encode individual constraints [14] and the creation of a watermarked solution that has a large percentage of values on the boundary of maximally tolerable infidelity.

4.6.2.4 De-synchronization

Traditional techniques for watermarking multimedia files that have a time dimension (e.g., audio and video) often require a synchronization signature encoder and signature verifier [17]. It is important to note that the proposed technique (as, for instance, can be seen from our demonstration examples) can be conducted on each time sample independently and can be applied to data that does not have a time dimension. For the case when a user decides to use a different set of constraints for each sample, one can just check which percentage of them is satisfied to establish the presence of the signature. This is because all imposed constraints are always selected (according to the particular digit in the signature) from the the same set of constraints. If a user decides to use an even more diverse set of constraints, she can add synchronization segments that consist of the same set of constraints followed by an ID that corresponds to the randomized part of the signature. Note that for verifying the signature in this case, one must store (that can be done on a desktop computer) all the IDs and their corresponding randomized components.

It is important to analyze how often and where each step of signature encoding and verification is conducted. The generation of the RSA algorithm is a single time event. Therefore, the majority of nodes just properly use the initialized RC4 cipher or some other block or stream cipher to generate the constraints. Hence, the computational and energy requirements are low.

4.6.3 Multi-modal Sensor Fusion: Data Preparation

We now present an approach for multi-modal sensor fusion. The approach starts by stating the multi-modal sensor fusion problem as a system of equations that is derived from the model of the physical world. Each measured variable is represented as the sum of its value and error of measurement. The final goal is to find unknown variables and our error corrections to measured variables so that the consistency of all equations and estimated errors is maximized. Numerous other approaches for multi-modal sensor fusion (MMSF) that use models such as Markov chains, cellular automata, interacting particles, and a variety of parametric and nonparametric statistical models have been proposed. Many more statistical models could be envisioned in the future.

There are two main reasons, in addition to brevity, why we limit our attention to the equation-based model. The first is that the process of solving equation with errors for consistency is very similar to the way other models are solved for consistency. Therefore, it is straightforward to adopt the proposed watermarking scheme to watermarking MMSF based on other models. More importantly, essentially all physical sciences and many life sciences overwhelmingly use equation-based models.

In this section, we formally state the equation consistency-based MMSF. We start by developing relevant equations from models of the physical world and measured phenomena. These equations have the following form:

$$\bar{F}(X_1, \ldots, X_m, Y_1, \ldots, Y_n) = 0 \qquad (4.11)$$

where X_i's are measured variables and Y_j's are calculated variables. Accounting for the errors, we can rewrite Equation (4.11) in the format shown in Equation (4.12).

$$\bar{F}(X_1 + \epsilon_1, \ldots, X_m + \epsilon_m, Y_1, \ldots, Y_n) \leq \delta_k \qquad (4.12)$$

where $k = 1, \ldots, p$. The final step is conversion of the equation into a canonical nonlinear programming form that consists of an objective function and constraints. Therefore, we convert Equation 4.12 to the following format:

$$OF \quad \min G(\epsilon_i, \delta_k)$$

$$s.t. \quad \bar{F}(X_1 + \epsilon_1, \ldots, X_m + \epsilon_m, Y_1, \ldots, Y_n) \leq \delta_k \qquad (4.13)$$

where $i = 1, \ldots, m$ and $k = 1, \ldots, p$. Now we are ready for the addition of watermarking constraints. Note that constraints can be added also to the objective function or can use new variables. The case when we only add p_1 watermarking constraints to Equation 4.13 will result in the following new constraint equations:

$$\bar{W}(X_1 + \epsilon_1, \ldots, X_m + \epsilon_m, Y_1, \ldots, Y_n) \leq \delta_k \qquad (4.14)$$

where $k = 1, \ldots, (p + p_1)$. The goal of watermarking is to add such constraints that the consistency of the solution is maintained, while one can establish the proof of the authorship by demonstrating that the obtained solution satisfies the set of additional (watermarking) constraints.

4.7 Driver Examples

We use two typical MMSF tasks as our driver examples: (1) acoustic atomic trilateration and (2) light characterization. The watermarking approach to acoustic atomic trilateration is described in Section 4.1 as the motivational example. In this section we present how the generic MMSF watermarking approach can be applied to light metering in sensor networks tasks.

We have two goals. The first is to vigorously specify all the key issues for light characterization watermarking. The second and probably much more important objective is to demonstrate that the application of generic active MMSF watermarking to numerous sensor network tasks is straightforward.

The light modeling equations (constraints) are of three types: (1) location of sensor node, (2) direction of the node with respect to angle of incoming light from a point light source, and (3) light intensity equations. The equations for location discovery are the same as in the case of acoustic atomic trilateration. The light intensity equations for each pair of light source–sensors have the form shown in Equation 4.15, where d is the distance between the source and the sensor, I_{src} is the intensity of the source, I_i is the measured intensity by the sensor, and $\cos\theta_i$ is the angle between the surface of the sensor and incoming light. We have as many equations of this type as there are pairs of light sources and sensors.

$$C_i : I_i d^2 - I_{src} \cos\theta_i = 0 \tag{4.15}$$

$\forall i = \{1, \ldots, k\}$, where k is the number of light sensors. After the inclusion of the 3D coordinates for the light source and the sensor, the equations of the type in 4.15 can be stated in the form:

$$I_i((x_{src} - x_i)^2 + (y_{src} - y_i)^2(z_{src} - z_i)^2) - I_{src} \cos\theta_i = 0 \tag{4.16}$$

The objective function is again the user-specific function of discrepancy in measurements and the degree to which the equations are not satisfying the light laws. The constraints for light intensity have the following form:

$$(I_i + \epsilon_i^I)((x_{src} - (x_i^m + \epsilon_i^x))^2 + (y_{src} - (y_i^m + \epsilon_i^y))^2$$

$$+(z_{src} - (z_i^m + \epsilon_i^z))^2) - I_{src}(\cos\theta_i^m + \epsilon_i^\theta) \le \delta_j \tag{4.17}$$

4.8   Experimental Results

The experimental section is organized in the following way. We first state the goals of our experiments, and then explain our simulation and experimental setup; finally we present data, data analysis, and try to extract main conclusions. The experiments are conducted on two driver examples: (1) acoustic atomic trilateration and (2) light source determination.

The primary goal of conducted experiments is to study the flexibility of the proposed watermarking techniques to the trade-off between the strength of the proof versus the watermarking overhead. The next goal is to study when and where each of the proposed watermarking techniques is most effective. Another important goal is to find the situations where one can conduct more aggressive watermarking without degrading the quality of results. These situations are characterized with respect to the level of error and the mutual relationship of the sensors. In addition to simulation on our three driver examples, we also conducted experimentation using the light compass appliance [21]. The goal is to check wether or not our models and our simulation setups are realistic. The first driver example is acoustic trilateration. The goal of acoustic atomic trilateration is to determine the position of a node, using measured distances between that node and k other nodes. In our experimentation, we use an average value of $k = 4.2$. We organized experiments into three different scenarios.

In the first, we placed all nodes randomly within a unit square. In the second, we placed the node that is the object of multilateration within a triangle formed by three other nodes that know their locations and we varied the angle between two beacon nodes between 1 degree and 179 degrees. In the final setup, we preset the nodes that know their position so that the lengths of edges between them have a given value of the Ginni coefficient. The Ginni coefficient indicates to what extent the length of the edges are equal. For example, a value close to 1 indicates that one edge is dominantly larger than two others. Value $1/3$ indicates that all three edges are of exact equal length. In all cases, we assume that both time and temperature measurements have errors with Gaussian distribution, with mean value zero and variance between 0.02 and 0.1. The trilateration is a very small example. Nevertheless, we were able to embed sufficiently long signatures that ensure low probability of coincidence while inducing only nominal overhead. Figure 4.6 shows the relationship between the strength of the authorship proof and the overhead with respect to a non-watermarked solution for different error values. From Figures 4.7 and 4.8 we see that high strength of the authorship can be achieved for a given value of overhead when the angle is close to 1 or 179 degrees or when the Ginni coefficient has a very high value. The explanation for this phenomenon is that for this situation, the non-watermarked solution has very high error and therefore it is very easy to hide additional information in it. This high error is a consequence of similarities between the equations used for trilaterations.

For the simulation with light sensors, we have developed a model of a room that has two to four point light sources placed on the ceiling and instrumented the room with 25 light sensors randomly placed and oriented. Each of the light sensors was exposed to at least one light; that is, we did not use sensors that were blocked or in the shade of each other or by themselves. The measurements for light intensity and distance used for location discovery had errors with a Gaussian distribution. The direction software compass error followed the error model presented in Priyantha et al. [23]. In Figure 4.9, we show the overhead versus the strength of authorship proof and the measurements error for the case of additional variables and constraints to the light equations.

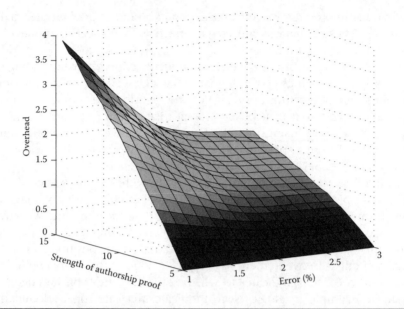

Figure 4.6 Acoustic multilateration location discovery. The *x*-axis indicates the strength of the authorship (probability of coincidence); shown value *x* on the diagram indicates the exponent in the expression 10^{-x}. The *y*-axis shows the level of error in the sensors. The *z*-axis indicates the average overhead for 200 measurements.

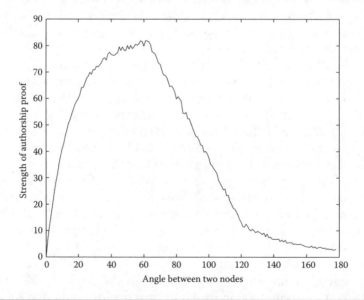

Figure 4.7 Acoustic multilateration location discovery. The normalized inverse values for the relative overhead for different constellations of the points. All values are normalized against the value for the triangle, where one angle is 1 degree.

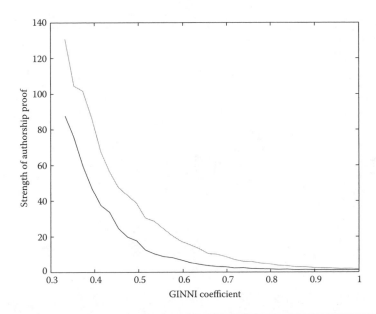

Figure 4.8   Acoustic multilateration location discovery. The normalized inverted values for the relative overhead for different constellations of points. All values are normalized against the value for the Ginni coefficient 0.99. The upper curve is for the level of 2% initial errors, and the lower curve is for the level of 1% for initial errors.

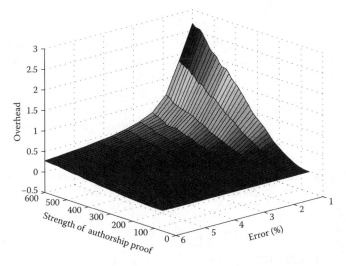

Figure 4.9   Light watermarking using addition of a new variable and constraints. The trade-off between the strength of authorship (x-axis) and the overhead (z-axis) for different values of sensor errors (y-axis). Shown value x on the diagram indicates the exponent in the expression 10^{-x}.

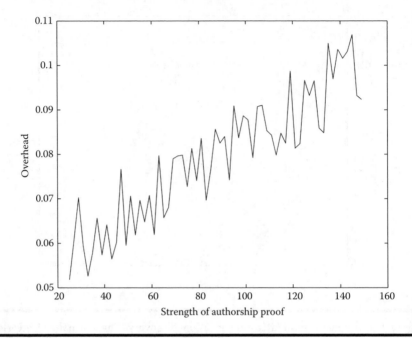

Figure 4.10 Experimental light appliance watermarking using addition of constraints. The trade-off between the strength of authorship (*x*-axis) and the overhead (*y*-axis). Shown value *x* on the diagram indicates the exponent in the expression 10^{-x}.

The last study was experimental. We positioned in a dark room a single point light and light compass equipped with five light sensors, as shown in Figure 4.2b. The trade-off between embedded watermark and strength of the proof is shown for the addition of constraint variables in Figure 4.10. Again, we see that very strong proof of authorship can be obtained at very minimal loss of accuracy. For all our simulations, we conducted 10,000 experiments for each scenario, but due to the time-consuming nature of experimentation, we conducted a total of 150 measurements. As a consequence, the trade-off graph has a significantly higher number of local minima that are most likely just artifacts of a relatively low number of experiments.

4.9   Conclusion

We have developed the first intellectual property protection (IPP) technique for data and information generated in sensor networks. The key idea is to impose additional conditions that correspond to the cryptographically encoded author/owner's signature. The technique is generic, optimization intensive, adaptive, transparent to sensor network middleware, and flexible. We validated the effectiveness of the approach on two canonic sensor network tasks: (1) location discovery and (2) light source characterization, using both simulation and prototyping. We achieved exceptional low probabilities of coincidence with only nominal quality degradation.

References

1. D. Boneh, J. Shaw. *Collusion-Secure Fingerprinting for Digital Data* CRYPTO, (1995), pp. 452–465.
2. J.T. Brassil, S. Low, N.F. Maxemchuk. Copyright Protection for the Electronic Distribution of Text Documents, *Proceedings of the IEEE*, Vol. 87 (No. 7) (1999), pp. 1181–1196.
3. P.T. Breuer, K.C. Lano. Creating specifications from code: Reverse engineering techniques, *Journal of Software Maintenance: Research and Practice*, 3, (1991), pp. 145–162.
4. S. Brin, J. Davis, H. Garcia-Molina. Copy detection mechanisms for digital documents, *Proceedings of the ACM SIGMOD Annual Conference*, (1995).
5. R.R. Brooks, S.S. Iyengar. *Multi–Sensor Fusion*, Prentice Hall, 1998.
6. C.-C. Chang, K.-F. Hwang, M.-S. Hwang. Robust authentication scheme for protecting copyrights of images and graphics, *IEEE Proceedings*, Vol. 149 (No. 1) (2002), pp. 43–50.
7. C. Collberg, C. Thomborson. Watermarking, Tamper-Proofing, and Obfuscation — Tools for Software Protection, University of Arizona Technical Report 2000–03 (2000).
8. V. Darmstaedter, J.-F. Delaigle, J.J. Quisquater, B. Macq. Low cost spatial watermarking, *Computers & Graphics*, Vol. 22 (No. 4) (1998), pp. 417–424.
9. W. Diffie, M. Hellman. New directions in cryptography, *IEEE Transactions on Information Theory*, Vol. IT-22 (No. 6) (Nov. 1976), pp. 644–54.
10. J. Fridrich, M. Goljan, A.C. Baldoza. *New Fragile Authentication Watermark for Images*. ICIP, (2000).
11. F. Hartung, M. Kutter. Multimedia watermarking techniques, *Proceedings of the IEEE*, Vol. 87 (No. 7) (July 1999), pp. 1079–1107.
12. J. Hightower, G. Borriello. Location systems for ubiquitous computing, *IEEE Computer*, Vol. 34 (No. 8) (2001), pp. 57–66.
13. J. Hill, R. Szewczyk, A. Woo, S. Hollar, D. Culler, K. Pister. System architecture directions for networked sensors, in *Proc. of the 9th Intl. Conf. on Architectural Support for Programming Languages and Operating Systems*, (Nov. 2000), pp. 93–104.
14. I. Hong, M. Potkonjak. Behavioral Synthesis Techniques for Intellectual Property Protection, *ACM/IEEE Design Automation Conference*, (1999), pp. 849–854.
15. N.F. Johnson, Z. Duric, S. Jajodia. *Information hiding: steganography and watermarking: attacks and countermeasures*, Boston: Kluwer Academic, (2001).
16. D. Kirovski, H.S. Malvar. Robust spread-spectrum watermarking, *IEEE International Conference on Acoustics, Speech, and Signal Processing*, (2001).

17. D. Kirovski, H. Attias. Audio watermark robustness to desynchronization via beat detection, *Information Hiding Workshop*, (2002).

18. F. Koushanfar, G. Qu, M. Potkonjak. Intellectual property metering, *Information Hiding Workshop*, (2001), pp. 87–102.

19. M. Kutter, F.A.P. Petitcolas. A fair benchmark for image watermarking systems, in *Proc. SPIE Security and Watermarking of Multimedia Contents*, Vol. 3657 (Jan. 1999), pp. 226–239.

20. J. Lach, W.H. Mangione-Smith, M. Potkonjak. Robust FPGA intellectual property through multiple small watermarks, *Design Automation Conference Proceedings*, (1999), pp. 831–836.

21. S. Meguerian, J.L. Wong, M. Potkonjak. Light Sensor Appliance: Techniques for Quantitative Design and Efficient Use, UCLA Technical Report, (October 2002).

22. A. Perrig, R. Szewczyk, V. Wen, D. Cullar, J.D. Tygar. SPINS: security protocols for sensor networks, *ACM Sigmobile (Mobicom)*, (July 2001).

23. N.B. Priyantha, A. Miu, H. Balakrishnan, S. Teller. The Cricket Compass for context-aware mobile applications, *ACM Sigmobile (Mobicom)*, (July 2001).

24. E. Praun, H. Hoppe, A. Finkelstein. Robust mesh watermarking, *Proc. Computer Graphics*, (1999), pp. 49–56.

25. G. Qu. Publicly Detectable Techniques for the Protection of Virtual Components, *IEEE/ACM DAC*, (2001).

26. V. Raghunathan, C. Schurgers, S. Park, M. Srivastava. Energy-aware wireless microsensor network, *IEEE Signal Processing Magazine*, Vol. 19 (No. 2) (March 2002), pp. 40–50.

27. J.P. Stern, G. Hachez, F. Koeune, J. Quisquater. Robust object watermarking: application to code, in A. Pfitzmann, Editor. *Information Hiding, Lectures Notes in Computer Science (LNCS)*, Vol. 1768, Dresden, (2000), pp. 368–378.

28. M.D. Swanson, B. Zhu, A. Tewfik. Transparent robust image watermarking, in *Proceedings of the IEEE Int. Conf. on Image Processing ICIP-96*, (September 1996), pp. 211–214.

29. M.G. Wagner. Robust watermarking of polygonal meshes, *Geometric Modeling and Processing*, (2000), pp. 201–208.

KEY MANAGEMENT

Chapter 5

Group Key Management in Sensor Networks

Sencun Zhu and Wensheng Zhang

Contents

5.1 Abstract

Broadcast is a typical communication model in a sensor network. For sensor networks deployed in adversarial environments, we must address the group rekeying problem, that is, how to update a group key shared by all the nodes in the network upon detection of a node compromise. In this chapter we present a centralized group rekeying scheme called GKMPSN and a distributed group rekeying scheme called PCGR. These two schemes offer different security and performance guarantees under different network models. Our analysis shows that our schemes have attractive and unique properties such as partial statelessness and localized node revocation.

5.2 Introduction

A sensor network is a wireless network consisting of many affordable and low-power sensor nodes that are capable of computation, storage, and communication. Because sensors can monitor and record information about their surroundings, sensor networks are expected to be extremely useful for a variety of environmental, civil, military, and homeland security applications, and will become part of our daily life.

Broadcast communication is an important communication pattern in a sensor network. To collect sensor data of interest, a network controller in a sensor network can broadcast missions, commands, or queries to all the nodes in the network. Broadcast communication is also used during sensor network maintenance. For example, after a large sensor network has been deployed, it is infeasible to gather the sensor nodes for reprogramming when the code in the sensor nodes has to be updated. In this case, we have to rely on the technique of reprogramming in the air [17], that is, broadcasting new code in the entire network through wireless channels.

For sensor networks deployed in often adversarial environments such as a battlefield, a secure communication model is needed to provide the confidentiality of broadcast communication. For example, the network controller must prevent his queries or missions from being understood by an adversary that is capable of eavesdropping the messages. To achieve this goal, an efficient solution is to let all nodes in the network share a key (called a group key) that is used by the network controller to encrypt broadcast messages and used by sensor nodes to decrypt broadcast messages.

Securing sensor networks, however, is complicated by the network scale, the highly constrained system resource, and the difficulty in dealing with node compromises. The low cost of sensor nodes (e.g., less than $1 as envisioned for smart dust [16]) precludes the built-in tamper-resistance capability of sensor nodes. Thus, the lack of tamper-resistance coupled with the unattended nature gives an adversary the opportunity to break into the captured sensor nodes to obtain the group key. To revoke the compromised nodes upon detection, the network controller will need to update the group key and distribute the new group key to all remaining nodes in a secure, reliable, and timely fashion. We refer to this process as *group rekeying.*

A simple approach for group rekeying is one based on unicast; that is, the network controller first encrypts the new group key with a secret key shared with an individual sensor node and then transmits the encrypted group key to that node. Despite its simplicity, this approach is not scalable because its communication cost increases linearly with the group size. For a group of size N, the network controller needs to encrypt and send N keys without considering packet losses. Moreover, considering the multihop nature of a large sensor network and the lack of knowledge in network topology, every encryption of the group key may have to be broadcast in the entire network. Clearly, this is not a scalable approach for resource-constrained sensor networks. In this chapter, we describe two scalable group key management schemes for sensor networks.

5.2.1 GKMPSN

The first one, called *GKMPSN*, is a centralized scheme in which a network controller broadcasts new group keys and node revocation information to all the nodes when detecting a node compromise. This scheme involves a *probabilistic key pre-distribution* phase in which each node pre-deploys a random set of keys out of a large pool of keys prior to the deployment of the network. A group rekeying operation involves two steps. In the first step, the predeployed keys are used for forming secure channels between neighboring nodes. These secure channels are then used for delivering new keying material to each legitimate sensor node. In the second step, each node uses the keying material it has received to independently update its group key as well as any predeployed keys that need to be changed because they were known to the compromised node(s).

In GKMPSN, a sensor node can decode the current group key even if it has missed a certain number of previous group rekeying operations. This property, which we refer to as *partial statelessness*, is extremely important in sensor networks where nodes may lose packets due to link errors or temporary network partitions. It also enables new sensor nodes that are added to the network after it has become operational to be able to securely communicate with other sensor nodes. In contrast to rekeying protocols developed for wired networks, the (per-node) communication cost of our protocol is independent of the group size and the number of nodes being revoked. Instead, the communication cost is a function of the network node density and the parameters of the probabilistic key predeployment scheme used by our protocol. The probabilistic nature of the underlying key predeployment scheme allows us to trade node storage for security and performance.

5.2.2 PCGR

We also present a second scheme called *PCGR*, which is a purely distributed group rekeying scheme. Different from previous research, PCGR is designed and implemented in a *distributed* manner, where no central authority is involved. This distributed property is critical for unattended sensor networks deployed in adversarial environments because the central authority is a single point of failure from security and performance perspectives.

PCGR is a periodic, local collaboration based group-rekeying scheme, which invalidates the group key known to a node being revoked. The basic idea is to let sensor nodes detect node compromises locally and update the group key periodically without involving a network controller. PCGR uses threshold secret sharing [15] to enable multiple nodes to collaboratively revoke a compromised node. Extensive analysis is conducted to evaluate the security level and the performance of the proposed schemes, as well as compare the performance of the proposed schemes with some existing group rekeying schemes. The analysis and simulation results show that the proposed schemes can achieve a good level of security and outperform most previously proposed schemes.

The remainder of this chapter is organized as follows. First we discuss some related work on key management in Section 5.3. Then we introduce GKMPSN in Section 5.4 and PCGR in Section 5.5. Section 5.6 concludes the chapter. For ease of presentation, we refer to the network controller as the *key server*.

5.3 Related Work

Recently, the group rekeying problem was extensively studied in the context of secure multicast in wired networks. Most of these schemes [1,10,12,18,21] use a logical key tree to reduce the complexity of a group rekeying operation from $O(N)$ to $O(logN)$. As in the unicast-based scheme, in all these schemes, the key server includes keys for all the member nodes when distributing its rekeying message, and every member receives the entire message although it is only interested in a small fraction of the content.

A recent effort [9] attempted to reduce the unnecessary keys a node has to receive by mapping the physical locations of the member nodes to the logical key tree in LKH [18] for a static sensor network. However, compared to the original LKH scheme, it is only possible to reduce the energy cost of a group rekeying by 15 to 37 percent, not mentioning that it may incur a larger overhead in some scenarios. Moreover, for this scheme to work, the key server must have global knowledge of the network topology.

In addition, the above schemes, when employed in sensor networks, raise a practical issue due to the small packet size [17]. Consider a group of $N = 1024$ nodes. To revoke a node, the number of keys to be distributed is 10 (assuming a binary key tree). Assume that every key is 10 bytes (8 bytes plus a 2-byte key id field), and that the payload of a packet is 29 bytes. The key server needs to broadcast 5 packets to avoid the fragmentation of keys. These packets must be *reliably* forwarded to all the nodes in a hop-by-hop fashion. This is a non-trivial task because of the unreliable transmission links and hidden terminal problems in wireless sensor networks.

Previous research on key management in sensor networks has focused on establishing pairwise keys between two nodes. There are schemes based on physical contact to bootstrap trust [14], schemes using a trusted third party (base station) [13], and schemes based on the framework of probabilistic key predeployment [3–5,8,23]. However, most of these schemes do not have any key updating mechanisms, and thus the security of these schemes could be greatly jeopardized when multiple compromised nodes collude to launch attacks.

In Zhu et al. [20], a key management framework called LEAP is presented. LEAP establishes four types of keys for each sensor node: an individual key shared with the base station, a pairwise key shared with another sensor node, a cluster key shared with multiple neighboring nodes, and a group key that is shared by all the nodes in the network. The design of LEAP is motivated by the observation that different types of messages exchanged between sensor nodes have different security requirements, and a single keying mechanism is not suitable for meeting these different security requirements. The group rekeying scheme in LEAP is based on hop-by-hop secure propagation of group keys using cluster keys as key encryption keys. Cluster keys are established through pairwise keys, and pairwise keys are established through an erasure-based scheme. Group rekeying in LEAP is extremely efficient because, on average, every node only transmits one key. GKMPSN is similar to LEAP in that it also adopts the approach of hop-by-hop secure propagation, but it uses the technique of probabilistic key sharing to establish pairwise keys between a pair of neighbor nodes.

For the environments of wired networks, some distributed schemes (e.g., Blundo's scheme [2]) have been proposed to allow a set of nodes to update a group key in a distributed way. In these schemes, each node contributes a random share to compute a new group key. These schemes are not scalable or efficient because each node must send its share to any other trusted nodes securely and reliably, and the cost increases rapidly as the group size increases. Our PCGR scheme differs from these schemes in that it does not require network-wide share exchange for key updating. Instead, only neighboring nodes need to collaborate to recover the new group keys that are preloaded before node deployment.

5.4   GKMPSN: A Centralized Group Rekey Scheme

GKMPSN exploits the property of a sensor network that member nodes are both hosts and routers. In IP Multicast, all group members are end hosts, and they have no responsibility for forwarding keying materials to other group members. In contrast, for group communication in a sensor network, the members of the group also act as routers. As such, in GKMPSN, the key server only has to deliver the new group key securely to the group members that are its immediate neighbors, and these neighbors then forward the new group key securely to their own neighboring members. In this way, a group key can be propagated to all the members. Because every node only needs to receive one encryption of the group key, the average transmission cost per node is one key independent of the group size.

For the above scheme to work, a fundamental requirement is the existence of a secure channel between every pair of neighboring nodes. GKMPSN provides secure channels through probabilistic key predeployment. The technique of probabilistic key predeployment has been applied in several studies [3,5,23]; however, to the best of our knowledge, none of these studies consider using this technique for group rekeying purposes. Below we describe how the basic scheme works; two extended schemes with improved security can be found in Zhu et al. [22].

5.4.1   Protocol Overview

Our group rekeying protocol involves a *key predistribution* phase and a *rekeying* phase.

■ **Key predistribution:** Prior to deployment of the sensor network, all nodes obtain a distinct subset of keys out of a large key pool from the key server, and these keys are used as key encryption keys (KEKs) for delivering group keys.

A rekeying operation itself involves three steps: authenticated revocation notification, secure group key distribution, and key updating.

■ **Authenticated node revocation:** When the key server decides to revoke a node, it broadcasts a revocation notice to the network in an authenticated way.

- **Secure key distribution:** The key server generates and distributes a new group key K. The key K is propagated to all the remaining nodes in a hop-by-hop fashion, secured with the noncompromised predeployed keys as KEKs.
- **Key updating:** After a node receives and verifies the group key K, it updates its own KEKs based on K.
- **Notation:** Below are the notations that appear in the remainder of this discussion.

  - u, v (in lower case) are principals such as communicating nodes.
  - R_u is a set of keys that u possesses, and I_u is the set of key ids corresponding to the keys in R_u.
  - I_C is the set of ids of the compromised keys known to the revoked nodes.
  - I_P is the set of ids of all the keys in the key pool P.
  - $\{f_k\}$ is a family of pseudo-random functions [6].
  - $\{s\}_k$ means encrypting message s with key k.
  - $MAC(k, s)$ is the MAC of message s using a symmetric key k.

5.4.2 Detailed Protocol Description

5.4.2.1 Key Predistribution

Each node is loaded with the following information:

1. Each node u is loaded with m distinct keys from the key pool P of l keys $\{k_1, k_2, \ldots, k_l\}$, and these keys are used as KEKs. A deterministic algorithm is used to decide the subset of keys R_u allocated to node u. Specifically, for each node, the algorithm generates m distinct integers between 1 and l using a uniform pseudo-random number generator upon the input of a node id. These integers are the ids of the keys for the node, and the node is loaded with the keys indexed by these ids. As a result, each key in the key pool has a probability of m/l to be chosen by each node. Note that this construction allows any node that knows another node's id u to determine I_u, the ids of the keys held by u.

2. Each node is loaded with the initial group key k_g that is used for securing group-wide communications, and an individual key that is only shared between the node and the key server.

3. Finally, each node is loaded with the commitment (i.e., the first key) of the key chain of the key server because we are employing μTESLA [13] for broadcast authentication.

Note that this key predistribution phase is equivalent to the member joining phase in traditional secure IP multicast. In this chapter, however, we are mainly concerned with group rekeying due to node revocations.

5.4.2.2 Authentic Node Revocation

To revoke a node, the key server broadcasts a notification to the network to initiate a group rekeying. The notification must be authenticated so that compromised nodes cannot revoke a legitimate node or spread malicious packets that could lead to inconsistency in our schemes.

We employ μTESLA [13] for broadcast authentication in our protocol due to its efficiency and tolerance to packet loss. μTESLA is based on the use of a one-way key chain along with delayed key disclosure. To use μTESLA, we assume that all the nodes and the key server are loosely time synchronized; that is, a node knows the upper bound on the time synchronization error with the key server.

Let u be the node being revoked and k_i^T be the to-be-disclosed μTESLA key. The key server updates $I_C = I_C \cup I_u$ (originally $I_C = \emptyset$). Let M be the id of the noncompromised key (i.e., keys in $I_P - I_C$) that is possessed by the maximum number of remaining nodes in the network. The key server generates the new group key as follows: $k_g' = f_{k_M}(k_g)$, and then broadcasts the following message:

$$Key\ Server \longrightarrow * : u, M, f_{k_g'}(0), MAC(k_i^T, u|M|f_{k_g'}(0)). \qquad (5.1)$$

We refer to $f_{k_g'}(0)$ as the *verification* key because it enables a node to verify the authenticity of the group key k_g' that it will receive later. The key server distributes the MAC key k_i^T after one μTESLA interval. A node receiving the above node revocation message and the MAC key K_i^T verifies the message using μTESLA. It will store the verification key $f_{k_g'}(0)$ temporarily if the verification is successful.

5.4.2.3 Secure Key Distribution

When a node u is revoked, all the keys it possesses, including k_g and the keys in R_u, must either be changed or discarded to prevent it from accessing future group communications. In our basic scheme, all the nodes except u update k_g while discarding all the keys in R_u.

When a node v receives the node revocation notice and the MAC key that arrives one μTESLA interval later, it verifies the authenticity of the notice based on μTESLA. If the verification is successful, node v deletes its keys that have ids in I_u. If node v possesses k_M, the key possessed by the maximum number of remaining nodes, it will generate the new group key $k_g' = f_{k_M}(k_g)$ on its own, as the key server does. Otherwise, it expects to receive k_g' from other nodes through a secure logical path.

To explain the key distribution process, we assume a multicast delivery tree rooted at the key server. In practice, the delivery tree should be already in place for the distribution of group data, and it can be constructed by any appropriate multicast/broadcast routing protocol. The key server initiates the process by sending k'_g to each of its children in the tree that does not have k_M through a *secure logical path* (discussed in detail below). A receiving node v can verify the key by computing and checking if $f_{k'_g}(0)$ is the same as the verification key it received earlier in the node revocation notice. The algorithm continues recursively down the delivery tree, that is, each node v that has received k'_g transmits k'_g to its own children that have no k_M via secure logical paths.

The probability that a particular key K is allocated to a node in the key predistribution phase is equal to m/l. Therefore, on average, the number of nodes that possess K is given by Nm/l for a group size of N. Recall that k_M is selected on the basis of being the key possessed by the maximum number of nodes. Thus, the number of nodes in the network that possess k_M is generally larger than Nm/l. Therefore, a fraction of nodes can compute the new group key independently without waiting for it to be delivered to them. To reduce the rekeying latency and increase the reliability of delivering the new group key, these nodes can also independently start propagating the new group key to their downstream neighbors in the multicast tree. Furthermore, the parents of these nodes do not need to transmit k'_g to these nodes, thus saving the energy involved in transmitting and receiving k'_g.

5.4.2.4 Logical Path Discovery

The *logical path discovery* process is necessary when a node wants to forward a new group key to its neighbors securely. We say there are logical paths between two nodes when (1) the two nodes share one or more keys (we call such paths *direct* paths); (2) the two nodes do not share any keys, but through other nodes they can transmit messages securely to each other (we call such paths *indirect* paths and call the involved nodes *proxies*).

In our design, it is very easy to find logical paths between two nodes. Because the key predistribution algorithm is public and deterministic, a node u can independently compute I_v, the set of key ids corresponding to a node v's key set. Therefore, without proactively exchanging the set of its key ids with others, a node knowing the ids of its neighbors can determine not only which neighbors share or do not share keys with it, but also which two neighbors share which keys. The latter knowledge is very valuable when node u does not share any keys with a neighbor node v, because node u can ask a neighbor (say x) which shares keys with each of them to act as a *proxy*. For example, suppose node u shares a key k_{ux} with node x, node v shares a key k_{vx} with node x, but no shared keys exist between node u and node v. To forward a new group key k'_g to node v, the following steps are taken:

$$u \longrightarrow x: \{k'_g\}_{k_{ux}}, \quad x \longrightarrow u: \{k'_g\}_{k_{xv}}, \quad u \longrightarrow v: \{k'_g\}_{k_{xv}}$$

From this example, we can see that a *proxy* node acts as a translator between nodes.

We call node x in the above example node u's *one-hop proxy* to v. More generally, node x is said to be node u's *i-hop proxy* if x is i hops away from u and x shares a key with both u and v. If u and v do not have any *direct* paths and indirect paths via a one-hop proxy to each other, they can resort to using a *multi-hop proxy*. Our protocol always uses any direct paths that exist between nodes in preference to indirect paths, as the use of an indirect path incurs additional computational and communication overhead.

5.4.3 Security Analysis

The security and correctness of the group rekeying scheme described above derive from the fact that (1) none of the revoked nodes can generate the new group key k_g' because they do not know k_M, (2) none of the revoked nodes can obtain the new group k_g' because it is transmitted via secure logical paths established with keys not known to any of the revoked members, and (3) every node can verify the new group key independently using the verification key it received in the authenticated revocation notice.

However, as the number of node revocations increases, the size of the set I_C also increases until, ultimately, $I_C = I_P$. That is, all the keys in P are known to the coalition of all the revoked nodes. Thus, there is no k_M that can be used to update k_g. Let $Pr(w)$ denote the covering probability that the collusion of w revoked nodes renders $I_C = I_P$, which is given by:

$$Pr(w) = \begin{cases} 0 & \text{if } w < l/m \\ (1 - (1 - \frac{m}{l})^w)^l & \text{if } w \geq l/m \end{cases} \tag{5.2}$$

For given l, m, and a covering probability p_0, we can calculate the value of w such that $Pr(w) = p_0$. For instance, for $l = 2000$, $m = 100$, and $p_0 = 90\%$, we have $w = 192$. That is, 192 nodes have a probability of 90 percent to cover the entire key pool.

Another security problem that arises as the number of revoked nodes increases is that the coalition of the revoked nodes may have keys that completely cover the key set of a legitimate node. Note that it is much easier for the compromised nodes to have keys to cover the m keys of a legitimate node instead of all the l keys in the key pool. Because compromised keys are discarded after node revocation, a legitimate node will no longer have any keys left that it can use to establish any logical paths to other nodes for obtaining the new group key. This node is therefore excluded from the network innocently, although it is still a legitimate node.

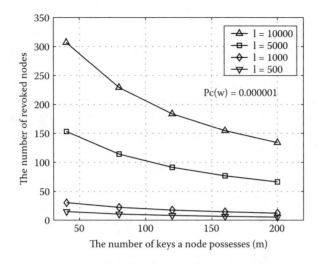

Figure 5.1 **The number of revoked colluding nodes (i.e., *w*) the scheme can tolerate under different *m* and *l* pairs, given $p_c(w) = 10^{-6}$.**

Let $p_c(w)$ be the probability that the m keys of a legitimate node are completely covered by that of w colluding revoked nodes. We have

$$p_c(w) = \left(1 - \left(1 - \frac{m}{l}\right)^w\right)^m \qquad (5.3)$$

From this equation, we know that by varying m and l we can obtain a desired security level. In Figure 5.1, we plot the number of colluding nodes (denoted as w_0) that the scheme can tolerate for a desired $p_c(w) = 10^{-6}$ for different m and l pairs. We observe that w_0 decreases with m but increases with l.

5.4.4 Communication Cost

The communication cost of a group rekeying operation is composed of the cost for broadcasting a node revocation notice and the cost for secure group key distribution. The revocation notice includes the id(s) of the node(s) being revoked, a verification key and a MAC of the notice, and the message is usually very small unless the number of nodes being revoked is very large.

In the secure key distribution phase of the rekey operation, a node transmits one encrypted key (i.e., the new group key) to each of its children in the delivery tree if they share a direct path. Every node receives one encrypted key; therefore, on average, every node transmits one encrypted key although an intermediate node may transmit more than one key, whereas a leaf node does not transmit at all. However, if a node does not have a direct logical path to a child node, it will need to communicate with a proxy node to form an indirect logical path. For a (round-trip) communication with a proxy node that is i hops away, the total number of keys transmitted is $2i$. Therefore, the communication cost for secure key distribution is determined by the numbers and types of logical paths existing between two neighboring nodes in the network, which are in turn determined by (1) the parameters l and m, (2) the network node density, and (3) the number of nodes being revoked at a batch. Here we only study the impact of probabilistic key sharing parameters. The reader is referred to Zhu et al. [22] for more details.

For a given l and m, p_s, the probability of directly sharing at least one key between any two nodes initially (i.e., prior to any node revocations), is

$$p_s = 1 - \frac{\binom{l}{m} \cdot \binom{l-m}{m}}{\binom{l}{m}^2} = 1 - \frac{\binom{l-m}{m}}{\binom{l}{m}} \tag{5.4}$$

Because the probability that a key is picked by both the nodes is $(\frac{m}{l})^2$, z_s, the expected number of keys shared between any two nodes initially (i.e., the number of *direct* paths between them), is $z_s = l \cdot (\frac{m}{l})^2 = \frac{m^2}{l}$.

From the above analysis, we see that for a given l, both p_s and z_s increase with m, while both p_s and z_s decrease with l for a given m. Also, we can see that m has a larger effect on z_s than l has. Thus, we can use the equations above to select l and m so that the probability of existing *direct* paths between two nodes is large enough. For example, to obtain $p_s = 99.5\%$, we can choose $l = 2000$ and $m = 100$ ($z_s \approx 5$). In this case, only 0.5 percent nodes have to receive the group key through an indirect path. Thus, the average transmission cost per node is close to one key.

From the earlier security analysis we observed that a smaller m and a larger l are more desirable from a security point of view; here we see that a larger m and a smaller l are more desirable from a performance standpoint. Thus, the scheme needs to make a trade-off between security and performance.

So far we have only shown the basic GKMPSN scheme and its rough performance. As can be seen, in the basic scheme, none of the preloaded keys are updated during a group rekeying process. Thus, it does not scale very well with the number of revoked nodes because the nodes revoked during different rekeying events may collude. This conclusion also applies to the previously probabilistic key predistribution schemes [3,5,23] because these schemes lack a key updating mechanism that prevents nodes compromised at different times from colluding. Zhu et al. [22] presented two improved schemes to address this issue, and also discussed various issues regarding transmission reliability and node additions. We have also implemented GKMPSN on a sensor network testbed, as a building block of our security for the Microsensor Networks project [11]. The source code is also available for download from the project Web site.

5.5 A Distributed Group Rekeying Scheme

To address the weakness of centralized node revocation schemes, we propose a periodic, local collaboration based group-rekeying scheme to invalidate the group key known to a node being revoked. The basic idea is to let sensor nodes update the group key periodically without using the key server. A simple solution for achieving this goal is to load every node with the same initial seed, the same pseudo-random function (PRF), and the same rekeying period. Periodically, every node obtains a pseudo-random number as the new group key based on the PRF. Because the rekeying period is the same, all nodes will have the same group key, and no online key server is required. This simple scheme, however, is insecure because a compromised node can compute all future group keys at once. It is a challenge to prevent a compromised node from deriving all future group keys by itself.

In our solution, a certain number of nodes collaborate to derive a new group key, under the assumption that there is a short time period T_s during which the node will not be compromised after its deployment. Specifically, every node is loaded with a polynomial for generating its future group keys. Using a threshold secret sharing scheme [15], the node divides the polynomial into multiple shares. The node sends one share to each of its neighbors and deletes the polynomial from its memory. Later it must receive a threshold number of shares from its neighbors to update the current group key. If the neighbors detect that this sensor node has been compromised, they can revoke this node simply by not sending their shares to the compromised node and not accepting/forwarding any of its packets. An attacker might reposition the compromised node to another location of the network, but the new neighbors do not possess the secret shares of this node. That is, once its neighbors stop providing secret shares, the compromised node will not be able to update the group key and is hence useless.

5.5.1 The Scheme in Detail

Our scheme runs in three steps, which are described in detail as follows:

1. The first step is *group key predistribution*. The key server generates a t-degree (t is a system parameter) univariate polynomial $g(x)$ over a prime finite field. This polynomial is for generating initial and future group keys, so we call it *group key polynomial* or g-polynomial. Let

$$g(x) = \sum_{0 \leq i \leq t} A_i x^i, \tag{5.5}$$

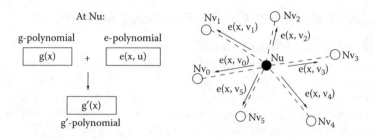

Figure 5.2 Node *u* encrypts the g-polynomial with a bivariate polynomial.

where x will be evaluated with key version number. $g(0)$ is the initial group key and $g(i)$ is the group key for the time interval i. Every node is loaded with $g(x)$ before it is deployed.

2. The second step is *polynomial share encryption and distribution*. A node, say *u*, randomly picks a polynomial $e(x, y)$. This polynomial is used for perturbing the original g-polynomial. The purpose is similar to encryption. So we call it *encryption polynomial* or e-polynomial. Let

$$e(x, y) = \sum_{0 \le i \le t, 0 \le j \le k} A_{i,j} x^i y^j \qquad (5.6)$$

where k is a system parameter, and x and y will be evaluated with key version numbers and node ids, respectively.

Node *u* perturbs $g(x)$ by adding it with $e(x, u)$, and the resulted polynomial (we call *encrypted group key polynomial* or g'-polynomial) is $g'(x) = g(x) + e(x, u)$. Suppose node *u* has n neighbors and their ids are v_is, where $i = 1, \ldots, n$. Node *u* divides $e(x, y)$ into n univariate polynomial shares $e(x, v_i)$ based on a (k, n) threshold secret sharing scheme, and then sends $e(x, v_i)$ to the neighbor v_i securely. Finally, node *u* erases $g(x)$, $e(x, y)$, and $e(x, u)$. This process is illustrated in Figure 5.2.

3. The third step is *key updating*. Every node keeps a rekeying timer that fires periodically and a key version count (denoted as c, initially set to 0) that is increased by one for each group rekeying. To help node *u* update the group key, a neighbor v_i increases c by one, derives $e(c, v_i)$ by evaluating its polynomial share $e(x, v_i)$ at $x = c$, and securely sends $e(c, v_i)$ to node *u*. Once receiving any k out of n shares, node *u* can reconstruct $e(c, y)$ based on interpolation. Finally, node *u* computes $e(c, u)$ by evaluating $y = u$ in $e(c, y)$ and computes $g'(c)$ by evaluating $x = c$ in $g'(x)$. This allows node *u* to derive the new group key $g(c) = g'(c) - e(c, u)$. This process is illustrated in Figure 5.3.

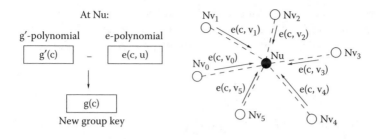

Figure 5.3 Node *u* recovers *e(c,u)*, an instance of the e-polynomial at time *c* (using the shares from neighbors), and then derives the group key *g(c)*.

For group key updating, a node needs to receive a certain number of polynomial shares from its trusted neighbors. The shares can only be used to calculate one instance of the e-polynomial that is necessary for computing the group key at a specific time point. A node cannot derive future group keys by itself. Thus, the node will be revoked from the network if its neighbors no long supply secret shares.

The selection of rekeying period shows a trade-off between security and performance. A larger rekeying period reduces the rekeying frequency and rekeying overhead, but increases the delay for invalidating the current group key. Nevertheless, the neighbors already know the compromise of the node; at least they can immediately start to drop messages from the compromised node without waiting until the next rekeying time point. This suggests that in practice a relatively large rekeying interval can be used.

5.5.2 Security Analysis

The security property of the PCGR scheme can be stated by Theorem 5.1.

Theorem 5.1

For a certain group, its g-polynomial $g(x)$ can be compromised if and only if:

    (1.1) *a node (N_u) of the group is compromised, and*
    (1.2) *at least $\mu + 1$ neighbors of N_u (In the encryption polynomial picked by N_u, i.e., $e(x, y)$, the degree of x is s and the degree of y is μ.) are compromised;*

or

    (2) *at least $s + 1$ past keys of the group are compromised.*

Proof (sketch), First, we prove that: if condition (1.1) and (1.2) or (2) is satisfied, $g(x)$ can be compromised.

Assume that N_u and its $\mu + 1$ neighbors [i.e., N_{v_i} ($i = 0, \ldots, \mu$)], have been compromised. The adversary can obtain $g'(x)$ and $e(x, v_i)$ ($i = 0, \cdots, \mu$). For an arbitrary x', a unique polynomial

$$e(x', y) = \sum_{j=0}^{\mu} B'_j y^j \tag{5.7}$$

can be reconstructed by solving the following $\mu + 1$ ($\mu + 1$)-variable linear equations:

$$\sum_{j=0}^{\mu} (v_i)^j B'_j = e(x', v_i), \text{ where } i = 0, \ldots, \mu. \tag{5.8}$$

Knowing $e(x', y)$, $g(x')$ can be computed as follows:

$$g(x') = g'(x') - e(x', u) \tag{5.9}$$

Similarly, we can prove that: if condition (2) is satisfied, $g(x)$ can be compromised.

Second, we prove that: if condition (1.1) is satisfied, but condition (1.2) and (2) are not satisfied, $g(x)$ cannot be compromised. Assume that N_u and its $m \le \mu$ neighbors [i.e., N_{v_i} ($i = 0, \ldots, m-1$)] are compromised. Thus, for an arbitrary x', the adversary can obtain $g'(x')$ and $e(x', v_i)$ ($i = 0, \ldots, m-1$). Since $m \le \mu$ and $e(x', y)$ is a μ degree polynomial of variable y, similar to the proof in Blundo et al. [2], we prove in the following that the adversary cannot find out $e(x', u)$:

We consider the worst case that $m = \mu$. Suppose the adversary guesses $e(x', u) = a$. A unique polynomial $e(x', y)$ (shown in Equation 5.7), which is a μ-degree polynomial of y, can be constructed by solving the following linear equations:

$$\begin{cases} \sum_{j=0}^{\mu} (v_i)^j B'^j = e(x', v_i), \quad i = 0, \ldots, \mu - 1 \\ \sum_{j=0}^{\mu} u^j B'^j = a. \end{cases} \tag{5.10}$$

However, because a can be an arbitrary value in $F(q)$, the adversary can construct q different polynomials $e(x', y)$. Also, because $e(x, y)$ is randomly chosen and hence $e(x', y)$ can be an arbitrary polynomial; that is, the q polynomials the adversary can construct are equally likely to be the actual $e(x', y)$. Formally,

$$\forall a, Pr(e(x', u) = a | \{e(x', v_i), 0 \le i \le \mu - 1\}) \\ \equiv Pr(e(x', u) = a) \tag{5.11}$$

Therefore, the adversary cannot derive $e(x', u)$.

Without knowing $e(x', u)$, the adversary cannot find out $g(x)$, which is equal to $g'(x) - e(x, u)$. On the other hand, because condition (2) is not satisfied, we assume that the adversary also knows $l \leq s$ keys of the considered group. Without loss of generality, we further assume that the compromised keys are $g(0), g(1), \ldots, g(l-1)$. Because $l \leq s$ and $g(x)$ is an s-degree polynomial of variable x, the adversary cannot find out $g(x')$ either.

Similar to the above proof, we can also prove that: if condition (1.2) is satisfied, but condition (1.1) and (2) are not satisfied, $g(x)$ cannot be compromised.

5.5.3 Performance Evaluation

5.5.3.1 Notations

We list some notations that are used in this section as follows:

- N: the total number of nodes in the network.
- n: the average number of trusted neighbors that a node has.
- n_c: the number of (compromised) nodes that should be evicted.
- L: the length (in bits) of a group key.

5.5.3.2 Performance Overhead of PCGR

In the PCGR scheme, each innocent node needs to send out n messages to its trusted neighbors during each key updating process. Each message includes one share, which has L bits. Therefore, nL bits are sent out by each node.

In terms of computational cost, each share sent/received by a node should be encrypted/decrypted using pairwise keys to prevent eavesdropping, and the total number of messages sent/received is $2n$ during each key updating process. Therefore, each node needs $2n$ encryptions/decryptions. In addition, each node N_u first needs to evaluate $n+1$ s-degree polynomials ($e_{v_i}(c)$ and $g'(x)$) to compute n shares and $g'(c)$, which needs $o((n+1)s^2)$ multiplications. After receiving $\mu + 1$ or more shares, it also needs to solve a $(\mu + 1)$-variable linear equation group to compute $e_u(c, u)$, and the computational complexity for solving such an equation group is $o(\mu^3)$ multiplications/divisions. The total computational complexity is $o((n+1)s^2 + \mu^3)$ multiplications/divisions over a finite field.

Regarding the storage cost, each node N_u needs to store the following information:

- The g'-polynomial $g'(x)$, which needs $(s+1) \cdot L$ bits to store its coefficients
- The shares of its neighbors' e-polynomials [i.e., $e_{v_i}(x, u)$ ($i = 0, \ldots, n-1$), which need $n(s+1)L$ bits]

The total storage requirement is $(n+1)(s+1)L$ bits.

Table 5.1 Comparing PCGR with Previous Group Rekeying Schemes

	SKDC	LKH	PCGR
Distributed	No	No	Yes
Data sent/received by each node (bits)	$\frac{(N-n_c)L}{2\sqrt{N}}$	$(2s_c - n_c)L$	nL
Rekeying delay	$o(\sqrt{N})$	$o(\sqrt{N})$	$o(1)$
Maximum computational overhead per node (at key updating time)	1 decryption	$logN$ decryptions	n decryptions, $o(\mu^3)$ multiplications/divisions over $F(q)$
Maximum computational overhead per node (overall)	1 decryption	$logN$ decryptions	$2n$ encryptions/decryptions, $o(\mu^3 + (n+1)s^2)$ multiplications/divisions over $F(q)$
Storage requirement per node (bits)	L	$LlogN$	$(n+1)(s+1)L$

5.5.3.3 Comparison with Other Group Rekeying Schemes

Table 5.1 compares PCGR with two previous schemes, SKDC [7] and LKH [18]. We only compare the costs of these schemes as related to key updating.

In the SKDC scheme, the central controller sends a new key to each trusted node individually. We assume that such a message should go through $\sqrt{N}/2$ hops on average, and each new key has L bits. Therefore, the total traffic introduced by key updating is $\frac{(N-n_c)\cdot L\cdot\sqrt{N}}{2}$, and the average size of the data sent/received by a node is $\frac{(N-n_c)L}{2\sqrt{N}}$. To finish a key updating, each node should receive the new key, which takes $o(\sqrt{N})$ units of time. The scheme is efficient in terms of computation and storage. Each node needs only one decryption and stores one key.

When analyzing the LKH scheme, we assume that a binary logic key hierarchy is used. Let s_c represent the size of the common ancestor tree (CAT) [1] of the evicted nodes. This scheme requires that each node on the CAT should change its key encryption key (KEK) and notify the KEK to its two children (except the evicted nodes). Therefore, $2s_c - n_c$ keys should be transmitted. Each node should receive the keys, which results in a rekeying delay of $o(\sqrt{N})$ units of time. Also, each node in this scheme should keep $logN$ (the height of tree) number of KEKs, and hence the storage requirement is $LlogN$.

Comparing our PCGR scheme to the previous schemes, we find that:

■ PCGR has smaller rekeying delay than the other schemes.
■ PCGR generates less traffic than SKDC and LHK.

■ As a distributed scheme, PCGR requires each node to perform some computations that are performed solely by the central controller in a centralized scheme. Therefore, the computational cost (per node) of PCGR is larger than SKDC and LHK. Note that in PCGR, only a small fraction of the computations should be performed at key updating time, so the rekeying delay is not increased very much. Furthermore, the key updating process can be initiated once every several hours or even a couple of days, so the computational cost is not significant in the long term.

■ The storage requirement of PCGR is larger than that of SKDC and LHK. However, the storage requirement is not very large in most cases. For example, if $n = 20$, $s = 30$, and $L = 64$ bits, the required storage space is about 5 kb. Note that, if the network density is very high and each node has many neighbors, the node can select only a subset of the neighbors to distribute shares. Thus, the storage requirement of each node can be reduced.

Note that the scheme introduced above does not answer the following questions. First, what if a node does not have neighbors? Second, what if more than k neighbors are compromised? The reader is referred to Zhang and Cao [19] for our solutions.

5.6 Conclusions

In this chapter we first introduced the broadcast (group) communication model in a typical sensor network. Then we studied the group rekeying problem for sensor networks deployed in adversarial environments. Finally we presented a centralized group rekeying scheme called GKMPSN and a distributed group rekeying scheme called PCGR. Security and performance analysis showed that our schemes have attractive properties, such as partial statelessness, localized node revocation.

References

1. D. Balenson, D. McGrew, and A. Sherman, Key Management for Large Dynamic Groups: One-Way Function Trees and Amortized Initialization. *IETF Internet Draft (work in progress)*, August 2000.
2. C. Blundo, A. Santis, A. Herzberg, S. Kutten, U. Vaccaro, and M. Yung, Perfectly-secure key distribution for dynamic conferences. In *Advances in Cryptology, Proceedings of CRYPTO'92*, 1993, LNCS 740, pp. 471–486.
3. H. Chan, A. Perrig, and D. Song, Random Key Predistribution Schemes for Sensor Networks. In *Proceedings of the IEEE Security and Privacy Symposim*, May 2003.
4. W. Du, J. Deng, Y. Han, and P. Varshney, A Pairwise Key Pre-distribution Scheme for Wireless Sensor Networks. In *Proceedings of 10th ACM*

Conference on Computer and Communications Security (CCS), Washington D.C., October 27–31, 2003.

5. L. Eschenauer and V. Gligor, A Key-Management Scheme for Distributed Sensor Networks. In *Proceedings of the 9th ACM Conference on Computer and Communications Security (CCS'02)*, 2002.

6. O. Goldreich, S. Goldwasser, and S. Micali, How to Construct Random Functions. *Journal of the ACM*, 33(4), 1986, 210–217.

7. H. Hugh, C. Muckenhirn, and T. Rivers, Group Key Management Protocol Architecture. *Request for Comments (RFC) 2093*, Internet Engineering Task Force, March 1997.

8. D. Liu and P. Ning, Establishing Pairwise Keys in Distributed Sensor Networks. In *Proceedings of the 10th ACM Conference on Computer and Communications Security (CCS '03)*, Washington D.C., October 2003.

9. L. Lazos and R. Poovendran, Energy-Aware Secure Multicast Communication in Ad-Hoc Networks using Geographic Location Information. In *Proceedings of IEEE ICASSP'03*, 2003.

10. D. Naor, M. Naor, and J. Lotspiech, Revocation and Tracing Schemes for Stateless Receivers. In *Advances in Cryptology — CRYPTO 2001*. Springer-Verlag Inc. LNCS 2139, 2001, 41–62.

11. URL: http://rogue.cs.gmu.edu/cs/index.html.

12. A. Perrig, D. Song, and D. Tygar, ELK, A New Protocol for Efficient Large-Group Key Distribution. In *Proceedings of IEEE Symposium on Security and Privacy*, 2001.

13. A. Perrig, R. Szewczyk, V. Wen, D. Culler, and J. Tygar, SPINS: Security Protocols for Sensor Networks. In *Proceedings of Seventh Annual ACM International Conference on Mobile Computing and Networks (Mobicom 2001)*, Rome Italy, July 2001.

14. F. Stajano and R. Anderson, The Resurrecting Duckling: Security Issues for Ad-Hoc Wireless Networks. In *7th International Workshop on Security Protocols*, Springer Verlag, 1999.

15. A. Shamir, How to Share a Secret. *Commun. ACM*, 22(11), 1979, 612–613.

16. The smartdust project, URL: http://robotics.eecs.berkeley.edu/pister/SmartDust/.

17. URL: http://www.tinyos.net/.

18. C. Wong, M. Gouda, and S. Lam, Secure Group Communication using Key Graphs. In *Proceeding of ACM SIGCOMM*, 1998, Vancouver, British Columbia, pp. 68–79.

19. W. Zhang and G. Cao, Group Rekeying for Filtering False Data in Sensor Networks: A Predistribution and Local Collaboration-Based Approach. *IEEE INFOCOM*, March 2005.

20. S. Zhu, S. Setia, and S. Jajodia, Leap: Efficient Security Mechanisms for Large-Scale Distributed Sensor Networks. In *Proceedings of the 10th ACM Conference on Computer and Communications Security (CCS '03)*, 2003, pp. 62–72.

21. S. Zhu, S. Setia, and S. Jajodia, Performance Optimizations for Group Key Management Scheme. In *Proceedings of 23rd International Conference on Distributed Computing Systems (ICDCS'03)*, 2003.

22. S. Zhu, S. Setia, S. Xu, and S. Jajodia, GKMPAN: An Efficient Group Key Management Protocol for Secure Multicast in Ad-Hoc Networks. In *Proceedings of the 1st International Conference on Mobile and Ubiquitous Systems (Mobiquitous'04)*, Boston, MA, 2004.
23. S. Zhu, S. Xu, S. Setia, and S. Jajodia, Establishing Pair-wise Keys for Secure Communication in Ad Hoc Networks: A Probabilistic Approach. In *Proceedings of the 11th IEEE International Conference on Network Protocols (ICNP'03)*, Atlanta, Georgia, November 4–7, 2003.

Chapter 6

Key Management Schemes for Wireless Sensor Networks

*Yixin Jiang, Chuang Lin, Minghui Shi, and
Xuemin (Sherman) Shen*

Contents

6.1 Abstract

Key management plays a very important role in establishing secure communications in wireless sensor networks (WSNs). This chapter presents a comphrehensive overview of the current state-of-the-art of key management for sensor networks: key distribution schemes, key agreement schemes, and key predistribution schemes (KPS). Our focus is on KPS, which may be the only feasible key management method for WSNs. The following aspects of KPS schemes are discussed: (1) shared key threshold KPS; (2) location-aware KPS; (3) pairwise key predistribution scheme with structured keypool; (4) path-key establishment KPS; (5) improved shared-key discovery KPSs, and their performance is evaluated. The chapter also discusses some open research issues in this area.

6.2 Introduction

Wireless sensor networks have emerged as an innovative class of networks with embedded systems [1, 2]. A WSN is a colloection of sensors whose size can range from a few hundred to a few hundred thousands sensors. The sensors do not rely on any predeployed network infrastructure; they thus communicate via an ad hoc wireless network. Sensors typically perform their tasks unattended, often in remote locations. They are usually specialized to monitor a specific environmental parameter such as thermal, optic, acoustic, seismic, or acceleration [3]. Thus, they may be deployed either inside, or nearby, target phenomenon to be studied.

Figure 6.1 shows a typical WSN architecture, which often contains one or more base stations that provide centralized control. A base station typically serves as the access point for users or as a gateway to another associated infrastructure such as data processing and management units. Individual sensors communicate locally with the neighboring sensors, and send their data over the peer-to-peer sensor network to the base station. Hence, there are three basic communication modes within WSNs: (1) node-to-node communication, (2) node-to-base station communication,

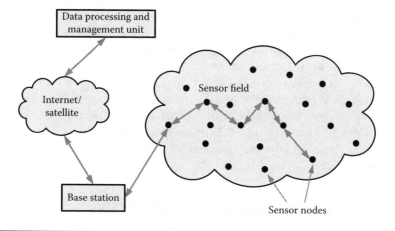

Figure 6.1   Architecture of wireless sensor network.

and (3) base station-to-node communication. A typical WSN is infrastructure-less, and deployed in various ways, usually via random scattering. The sensors are low-cost, highly power constrained, and limited in computation, storage capacity, communication bandwidth, and radio transmission power. Such constraints require serious consideration in designing the WSNs.

WSNs are often deployed in open and hostile environments, and thus are subjected to great security risks. Establishing secure communications is an open, difficult issue for WSNs, which involves the key setup and distribution. Managing secret keys is one of the prime security requirements in any network. Currently, there exist three key management schemes: (1) ***key distribution scheme***, (2) ***key agreement scheme***, and (3) ***key predistribution scheme***. The traditional key distribution scheme requires a trusted server to establish shared session keys between nodes and is prone to directed attacks against this central point. The key agreement scheme depends upon asymmetric cryptography protocols and algorithms. However, with the low resource constraints of sensors, asymmetric cryptography limits the practical use of this scheme. Presently, the only practical scheme for key management in large sensor networks may be key predistribution scheme, where key information is installed in each sensor node prior to deployment. Typically, there are two naive solutions for implementing secure pairwise communication: (1) a unique session key carried by all nodes, or (2) each sensor stores a set of separate $n-1$ pairwise keys, each of which is secretly shared with another sensor. Both are inadequately used in WSNs, because stealing the unique session key may compromise the entire network and storage of $n-1$ pairwaise keys in each sensor bounds practical deployments. To overcome the limitations, many feasible key management schemes for WSN have been proposed.

In this chapter we review the current state-of-the-art of key management schemes for WSNs, and compare them with respect to several evaluation metrics. We also attempt to provide insight into future trends in the area and outline the approaches that are likely to play a major role. The remainder of this chapter is organized as follows. In Section 6.3 we outline specific issues to be considered when designing key management schemes for WSNs. We review the existing three types of key management schemes for WSNs in Sections 6.4 and 6.5, respectively. In Section 6.6 we discuss some open issues in key management for WSNs, followed by the conclusions in Section 6.7.

6.3 Issues of Key Management in WSNs

Offering efficient key management in WSNs faces many challenges that will significantly impact the design and implementation of security protocols for WSNs. The majority of WSN applications are in areas where rapid deployment and dynamic configuration are necessary. Generally, the following constraints in WSNs should be carefully considered [4,5]:

1. Lack of fixed infrastructure.
2. Very large and dense deployment.
3. Vulnerability of nodes to physical capture: because sensors can be deployed in public and hostile locations, this exposes sensor nodes to physical attacks by an adversary, which may undetectably capture a sensor node and compromise the secret keys.
4. Vulnerability due to the over-reliance on base stations: the base stations in WSNs are centralized, powerful, and expensive. Hence, it is tempting to rely on them too much in the functions. This may attract attack on the base station and limit the application of the security protocol.
5. Lack of a prior deployment knowledge: usually, a WSN is deployed in a random way, and the sensors cannot know the exact node location information in advance. Hence, designing a security scheme should not assume the exact deployment knowledge of nodes.
6. Impaired wireles channel and limited bandwidth: the security protocol for WSNs should be designed to minimize message exchanges and message size.
7. Limited memory resources: the amount of storage memory in each sensor is constrained so that it is unsuitable to establish pairwise keys with each sensor in WSNs.
8. Limited computation capacity: the limited computation in nodes makes them vulnerable to denial-of-service (DoS) attacks [6,7].

Such limitations render it undesirable to use a public key cryptography algorithm, such as Diffie-Hellman [8] or RSA schemes [9]. Therefore, compared with conventional wired computing context, sensor networks have many characteristics that make them more vulnerable to attack. Some special security and performance requirements should be emphasized in WSNs:

1. *Resilience against node capture*: an adversary can mount a physical attack on a sensor node after the deployment. It is required to estimate the fraction of total network communications that are compromised by such captured nodes.

2. *Resilience against node replication*: an attacker can insert additional hostile nodes into a WSN. This is a serious attack because even a single compromised node might allow an adversary to populate the network with the clone of the captured node to such an extent that the illegitimate nodes could be over-numbered and the adversary can thus gain full control of the network.

3. *Node revocation or participation*: a new sensor should be dynamically deployed in a WSN, and a detected misbehaving node should also be dynamically removed from the system.

4. *Scalability*: when the number of sensors grows, the security may be weakened. It is neccessary to explore the maximum supported network size for a given deployment policy, as different key deployment policies will result in different network scales, which heavily impacts the scalability of the key schemes.

The first two requirements mainly concern network robustness; when some sensor nodes conspire or be compromised by an adversary, the keys of the other nodes are still secure. In addition, when establishing a pairwise key for any two sensors, the Byzantine attacks [10] must be considered.

6.4 Key Management Schemes

The three classes of key management — *key distribution scheme, key agreement scheme,* and *key predistribution scheme* (KPS) — have their own limitations, when applied to sensor network.

The traditional **key distribution schemes** [11,12] rely on online trusted third parties (TTP) to distribute session keys to nodes. These schemes are infeasible for WSNs, because the TTP may be out of reach or not available to some of the nodes due to communication range limitations, node movements, network dynamics, and unknown topology prior to development. A number of schemes, which were proposed for ad hoc networks [13–15], attempt to increase the availability of the key distribution service by replacing the online TTP with a subset of nodes organized arbitrarily or hierarchically. However, the performance of these schemes, in terms of efficiency and scalability, still depends on TTP; compromising the TTP affects all the distributed keys.

A *key agreement scheme* [16–24] is a self-enforcing scheme, in which each node takes part in establishing a shared secret through mutual exchanged messages among the nodes in a secure manner. Although these protocols are nearly fully distributed or self-organized without needing TTP, they are still not infeasible in WSNs for the following reasons. First, these protocols are not robust to variable topology or intermittent links frequently occurring in WSNs. To successfully establish a shared key, the underlying networks are strictly required to either support broadcasting or have a relatively time-invariant network topology, with the result that a routing infrastructure is requisite because pairwise nodes may not reach each other and need intermediate nodes to relay the messages. In addition, all nodes need to be online before the key agreement process is over; if any node leaves in the midst due to link or battery outage, the remaining nodes need to re-run the process from scratch. Evidently, these requirements could not be satisfied in WSNs. Second, a key agreement scheme that depends on asymmetric cryptography is not computationally efficient, as the limited computation and memory capacity of sensor nodes make such public-key cryptography operations impractical. Hence, the group key agreement approach without further improvement is not practical for WSNs. Finally, due to frequent interactive re-keying, the scalability in a key agreement scheme may be troublesome.

Some recent works demonstrate that the *key predistribution schemes*, independently introduced by Blom [25] and Matsumoto et al. [26], offer practical and efficient solutions to the key management problem in sensor networks [4,27]. In KPS, an offline TTP preinitializes each node in a set S with some secret information by which any subset of node $S_i \subseteq S$ later can find or compute noninteractively a common session key. Two works on KPS for distributed sensor networks [4,27] further suggest that KPS may be the only practical option for key management in WSNs, in which network topology is unknown prior to the deployment or changes fast after deployment. Therefore, our focus here is on KPS.

6.5 Key Predistribution Schemes

To improve the resilience to capture attacks on key management schemes for a large-scale WSN, many KPS have been proposed. In these schemes, a set of keys is randomly selected from a key pool and installed in the memory of each sensor. The pure Random KPS (R-KPS) was first proposed by Eschenauer and Gilgor [27]. Based on the basic R-KPS, some schemes are proposed to further improve its performance and security in the following five aspects:

1. *Shared-key threshold*: the q-composite random KPS [4] increases the amount of key overlap required for key-setup, it improves the

resilience against node capture attack, in which attackers can capture sensors and derive the preinstalled key information.

2. ***Location-aware R-KPS***: the known deployment information of sensors is used in key predistribution and sensor deployment [28–30], which also targets at enhancing the resilience to node capture.

3. ***Pair-wise key predistribution scheme with structured key pool***: structured key-pool KPS [31–33] is mainly used in pairwise key predistribution scheme, which can efficiently improve the resilience to node capture attack.

4. ***Path-key establishment scheme***: multi-path key establishment [4,34,35] can prevent a few compromised sensors from knowing the established pairwise keys.

5. ***Improved shared-key discovery*** [36,37]: efficient key discovery algorithm can minimize energy consumption by decreasing the message exchanges or computation complexity. The one-way function schemes [36,37] not only reduce the communication overhead during the key discovery phase, but also improve the resilience to node fabrication attack, in which attackers can fabricate new nodes based on information derived from the captured nodes.

6.5.1 Basic Random KPS

Eschenauer and Gilgor [27] first proposed a Random Key Predistribution Scheme (R-KPS) for WSNs, called basic scheme or EG scheme. It is based on probabilistic key sharing and a simple shared-key discovery protocol for key distribution, key revocation, and node re-keying. In the EG scheme, prior to sensor network deployment, each node randomly picks a subset of keys (called key ring) from a large key pool. Upon deployment, any pair of nodes can establish a secure and direct connection if they share at least one common key. Because the keys are randomly distributed to each sensor node, not all pair of nodes have common keys. Hence, any two nodes without common keys are required to establish an indirect connection via an intermediary node that shares a common key with both nodes, respectively. More specifically, R-KPS consists of three different phases: *key predistribution, shared key discovery,* and *path-key establishment.*

The ***key predistribution phase*** takes place before the sensor nodes are deployed. During this phase, a large pool of random keys and their key identifiers are generated. A subset of randomly chosen keys and their associated key identifiers are assigned to each sensor. This phase is to guarantee that, with a small number of keys, a common key is established with high probability between two or more sensors during the shared key discovery phase.

The ***shared-key discovery phase*** performs during WSN initialization. Each node discovers its neighbors with which it shares keys within the wireless communication range. A simple way for any two nodes to discover if they share a key is that each node broadcasts the list of identifiers of the keys in their key ring in cleartext. A more secure method is that, for each key on a key ring, each node broadcasts $\{\alpha, E_{K_i} = (\alpha), i = 1, 2, \ldots, k\}$, where α is a challenge. Once $E_{K_i}(\alpha)$ is decrypted with a proper key by a recipient, the challenge is revealed and thus a common key shared between the broadcasting node and the recipient is established. This phase can establish a direct secure link between two nodes if they share a key.

The **path-key discovery phase** executes after the shared-key discovery phase. In this phase, a path-key is established for any selected sensor pairs that do not share a common key but can be connected by two or more direct links created in the shared-key discovery phase. If the key graph is connected, a indirect link can be found from a source node to the target node. The source node can then generate a path-key and send it securely via the path to the target node.

In the EG scheme, a key issue is to pick the right parameters such that the graph generated during the key-setup phase is connected. Consider a random graph $G(n, p)$, where n is the number of nodes and p denotes the probability that a link exists between any two nodes. When $p = 0$, the graph does not have an edge; whereas when $p = 1$, the graph is fully connected. Erdos and Renyi [38,39] showed that there exists a value p_i such that, in a very large random graph, the connectivity property moves from "non-existent" to "certainly true." Hence, it is possible to calculate the expected degree d for the vertices in the graph such that the graph is connected with a high probability P_c. Eschenauer and Gilgor [27] calculated the requisite expected node degree, in terms of the size n of the network, as $d = \left(\frac{n-1}{n}\right)[\ln n - \ln(-\ln P_c)]$.

For a given deployment density of WSNs, let $n' \ll n$ be the expected number of neighbors within communication range of a node; then the probability of sharing at least one key between any two nodes in a neighborhood is $p' = d/n'$. Assume that the size of key rings is k and the key pool is P. For a given p', P can be derived as a function of k:

$$p' = 1 - \frac{((P-k)!)^2}{(P-2k)!\,P!} \simeq \frac{(1-k/P)^{2(P-k+0.5)}}{(1-2k/P)^{(P-2k+0.5)}}$$

A fundamental problem in secure WSNs is to choose the relevant parameters, pool size, and key ring size, to achieve a very high probability of connectivity. This problem was earlier addressed by Eschenauer and Gilgor [27]. However, Pietro et al. [40] point out that the model in [27] is not completely satisfactory for secure WSNs, because this model does not consider the geometric position of the sensors in space and, as a consequence, it cannot properly model the inherent locality of physical visibility among sensors. The Erdos-Renyi model [38,39] assumes that edges exist independently. As shown in Pietro et al. [40], this is far from being true and thus the Erdos-Renyi model cannot be used as a reliable guide to design secure WSNs. Pietro et al. precisely indicate how to choose the proper pool size and key ring size to guarantee very high probability of connectivity. Other similar literature can be seen in Hwang and Kim [41], Pietro et al. [42], and Xue and Kumar [43].

6.5.2 Shared-Key Threshold R-KPS: The q-Composite R-KPS

The q-Composite R-KPS [4] is a modification of the basic scheme, differing only in the size of the key pool S and using multiple keys to establish communication instead of just one. By increasing the amount of key overlap required for key-setup, the resilience capability of the network against node capture is strengthened.

In the initialization phase, a set S of random keys is picked from the total key space; m random keys are selected from S and are stored into the node key ring for each node in the network. The nodes perform key discovery with their neighbors before each node can identify every neighbor node with which it shares at least q keys. Let the number of actual keys shared be q', where $q' > q$. A new communication link key K is generated as the hash of all shared keys, for example, $K = \text{hash}(k_1 \parallel k_2 \parallel \ldots \parallel k_{q'})$.

Let m be the number of keys that any node can hold in its key ring, and let $p(i)$ be the probability that any two nodes have exactly i keys in common. Then, the probability of any two nodes sharing sufficient keys to form a secure connection can be computed as

$$P_{conn} = 1 - \sum_{i=1}^{q-1} \frac{\binom{|S|}{i} \binom{|S|-i}{2(m-i)} \binom{2(m-i)}{m-i}}{\binom{|S|}{m}^2}$$

In this scheme, the size of the key pool, $|S|$, is a critical parameter. If $|S|$ is too large, then the probability of any two nodes sharing at least q keys would be less than p, and the network may not be connected after the bootstrap is complete. If $|S|$ is too small, the security strength may be weakened. Hence, for a given key ring size m, minimum key overlap q, and minimum connection probability p, the largest $|S|$ is chosen such that $p_{conn} > p$.

Assuming that the number of captured nodes is x and each node contains m keys, the probability that any secure link setup in the key-setup phase between two un-compromised nodes is compromised, or the resilience capability against node capture, is estimated as

$$\sum_{i=q}^{m}\left[1-\left(1-\frac{m}{|S|}\right)^{x}\right]^{i}\cdot\frac{p(i)}{p}=\sum_{i=q}^{m}\left[1-\left(1-\frac{m}{|S|}\right)^{x}\right]^{i}\cdot\frac{1}{p}$$

$$\cdot\frac{\dbinom{|S|}{i}\dbinom{|S|-i}{2(m-i)}\dbinom{2(m-i)}{m-i}}{\dbinom{|S|}{m}^{2}}$$

The q-Composite R-KPS offers greater resilience against node capture when the number of nodes captured is small. When a large number of nodes has been compromised, the q-Composite R-KPS tends to reveal large fractions of the network to the adversary. Increasing q makes it more difficult for an adversary to obtain small amounts of initial information from the network via a small number of initial node captures. It comes at the cost of making the network more vulnerable once a large number of nodes has been breached. This may be a desirable trade-off because small attacks are cheaper to mount and much more difficult to be detected than large-scale attacks.

6.5.3 Location-Aware R-KPS

In many practical applications, the deployment location of sensors may be available *a priori*. Knowing which sensors are close to each other is helpful for key predistribution. In WSNs, the primary goal of secure communication is to provide such communications among neighboring nodes. Hence, the most important knowledge that can benefit a key predistribution scheme is the deployment knowledge of the nodes that are likely to be the neighbors. Once the neighbors of each sensor can be deterministically known, key predistribution becomes easy. However, owing to the random deployment of nodes, it is unrealistic to know the exact set of neighbors of each node, but knowing the set of possible or likely neighbors for each node is much more feasible. That is, it is usually possible to approximately determine their locations, although it is difficult to precisely pinpoint the sensor's positions in static sensor networks.

6.5.3.1 Location-Aware R-KPS: Liu and Ning's Scheme

By taking advantage of the sensors' expected locations, Liu et al. [28] first presented a simple location-aware deployment model and proposed two new R-KPS, *closest* R-KPS and *location-based* R-KPS using bivariate polynomials.

The **closest R-KPS** predistributes pair-wise keys for pairs of sensors, so that any two sensors would share a common pair-wise key if both of them appear in each other's signal range with a high probability. However, the scheme is unpractical, because it is not easy to get the exact deployment probability that two sensors are neighbors. In the key predistribution phase, the setup server will, using the expected locations of the sensors, predistribute keys for pairs of sensors to establish their pair-wise key. Specifically, for each sensor a, the setup server first discovers a set S of other sensors whose expected locations are closest to the expected location of a. For each sensor b in S, the setup server randomly generates a unique pair-wise key $K_{a,b}$ if no pair-wise key between a and b has been assigned. The setup server then distributes $(b, K_{a,b})$ and $(a, K_{a,b})$ to sensors a and b, respectively. The other two phases, the shared key discovery phase and path-key establishment phase, are similar to those in the basic R-KPS.

Evidently, the closest R-KPS can achieve better performance if the location information is available. However, the closest R-KPS still has some limitations. Given the constraints on the storage capacity, sensor density, signal range, and deployment error, the probability of establishing a direct key is fixed. For a special WSN, it is not convenient to adjust the last three parameters. So, to increase the probability of establishing the direct key, the storage capacity of sensor nodes needs to increase for pair-wise keys.

To address these limitations, Liu et al. developed another scheme, *location-based R-KPS*, using bivariate polynomials and the location information. The scheme makes a trade-off between the security against node captures and the probability of establishing direct keys under the constraint of a given memory. It does not require the setup server be aware of the global network topology, thus makes the deployment easier.

The main idea of the location-based R-KPS is to combine the closest R-KPS with the polynomial-based key predistribution technique. Specifically, as shown in Figure 6.2, the target field is partitioned into small areas called *cells*. Each cell is associated with a unique random bivariate polynomial. Unlike the closest R-KPS, instead of assigning the pair-wise keys to each sensor and its closest sensors, the setup server distributes to each sensor a set of polynomial shares. These polynomial shares belong to the cells closest to the one that is the expected location of this sensor.

In location-based R-KPS, the probability of establishing a common key directly between any two neighboring nodes can be estimated by

$$p = \frac{\sum_{C_{j_c, j_r} \in S_{i_c, i_r}} p(C_{j_c, j_r}, C_{i_c, i_r})}{\sum_{C_{j_c, j_r}} p(C_{j_c, j_r}, C_{i_c, i_r})}$$

where $p(C_{j_c, j_r}, C_{i_c, i_r})$ is the probability of establishing a common key between any two nodes a and b, which are expected to be deployed in cell

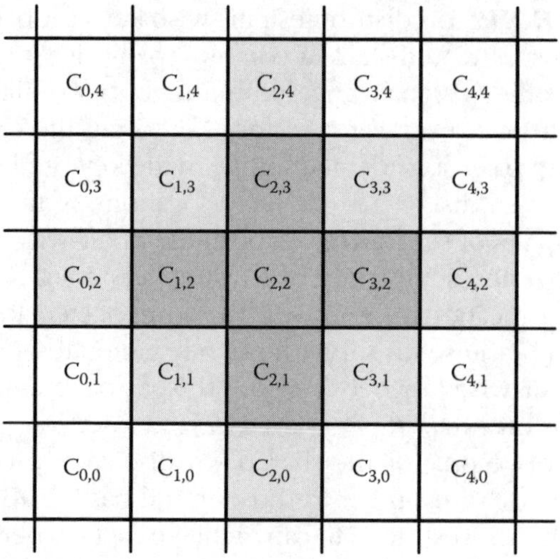

$C_{0,4}$	$C_{1,4}$	$C_{2,4}$	$C_{3,4}$	$C_{4,4}$
$C_{0,3}$	$C_{1,3}$	$C_{2,3}$	$C_{3,3}$	$C_{4,3}$
$C_{0,2}$	$C_{1,2}$	$C_{2,2}$	$C_{3,2}$	$C_{4,2}$
$C_{0,1}$	$C_{1,1}$	$C_{2,1}$	$C_{3,1}$	$C_{4,1}$
$C_{0,0}$	$C_{1,0}$	$C_{2,0}$	$C_{3,0}$	$C_{4,0}$

Figure 6.2   Partition of a target field.

C_{j_c, j_r} and C_{i_c, i_r}, respectively. S_{i_c, i_r} denotes the set of home cells of the sensors that share at least one common polynomial with a sensor whose home cell is C_{i_c, i_r}.

According to the analysis result of the polynomial-based KPS, as long as no more than t polynomial shares of a bivariate polynomial are exposed, an attacker knows nothing about the noncompromised pair-wise keys established by this polynomial. Assume that each sensor has the probability p_c of being compromised. Thus, among N_S sensors that share the polynomials of a particular cell, the probability that exactly i sensors have been compromised can be

$$P_c(i) = \frac{N_S!}{(N_S - i)!i!} p_c^i (1 - p_c)^{N_S - i}$$

Accordingly, the probability that the bivariate polynomial assigned to this cell is compromised is $P_c = 1 - \sum_{i=0}^{t} P_c(i)$. For any pair-wise key established directly between noncompromised sensors, the probability that it is compromised is the same as P_c.

6.5.3.2   Location-Aware R-KPS: Group-Based Scheme

Based on the R-KPS, Du et al. [29] also proposed a novel scheme by exploiting the node deployment knowledge such that the probability to find a common secret key between any two neighboring nodes can be maximized while other performance metrics are not degraded.

Deployment knowledge can be described using probability density functions (*pdf*). When the *pdf* is uniform, no information can be obtained on where a node is more likely to reside. The basic R-KPS assumes such a uniform distribution, whereas Du et al.'s KPS uses a nonuniform *pdf.* To demonstrate the effectiveness of this method, a specific distribution, the normal (Gaussian) distribution, is explored in depth. The related results show substantial enhancement over existing R-KPS, especially improving the connectivity of a network and resilience to node capture, thus reducing the required memory in sensor nodes.

To model the deployment knowledge, Du et al. first assume that the sensor nodes are static once they are deployed, and the target deployment area is a two-dimensional rectangular region with size $X \times Y$. The *pdf* for the location of node i, $i = 1, 2, \ldots, N$, over the two-dimensional region is given by $f_i(x, y)$, where $x \in [0, X]$ and $y \in [0, Y]$. With this general model, Du et al. proposed a *group-based deployment model.* The basic R-KPS for sensor networks is a special case: all sensor nodes are uniformly distributed over the whole deployment region.

In the *group-based deployment model,* N nodes are divided into $t \times n$ equal-sized groups so that each group $G_{i,j}$ ($i = 1, \ldots, t; j = 1, \ldots, n$) is deployed from the deployment point x_i, y_j, which are arranged in a grid. During the deployment phase, resident points of node k in group $G_{i,j}$ satisfy the *pdf* $f_k^{ij}((x, y) \mid k \in G_{i,j}) = f(x - x_i, y - y_j)$. If the deployment distribution follows a two-dimensional Gaussian distribution, for the deployment point (x_i, y_j) of group $G_{i,j}$, the *pdf* for node k is denoted as

$$f_k^{ij}((x, y) \mid k \in G_{i,j}) = \frac{1}{2\pi\delta^2} e^{-[(x-x_i)^2+(y-y_i)^2]/2\delta^2} = f(x - x_i, y - y_j)$$

where $f(x, y) = \frac{1}{2\pi\delta^2} e^{-x^2+y^2/2\delta^2}$. If a node is selected to be a given group with equal probability, the average deployment distribution (*pdf*) of any node over the entire region is $f_{overall} = \sum_{i=1}^{t} \sum_{j=1}^{n} \frac{1}{nt} f_k^{ij}((x, y) \mid k \in G_{i,j})$.

Compared with the basic R-KPS, the group-based deployment scheme offers nice local connectivity and global connectivity. The local connectivity is substantially improved. In addition, due to the use of deployment knowledge, the security properties, such as resilience to node capture, are also improved because the number of unnecessary keys carried by each sensor node is reduced.

6.5.3.3 Location-Aware R-KPS: Grid Group-Based Scheme

Although the proposed location-aware KPS using deployment information (node location information) in Liu and Ning [28] and Du et al. [29] can improve the resilience to node capture attack, in practice, the open or hostile deployment environment of sensor networks makes it easier for attackers to locate and selectively capture sensors, which can provide more information for the attacker to intrude the sensor network. In addition, due to the lack of node authentication, attackers can fabricate nodes using the secrets preinstalled in the captured nodes.

To address these issues, Huang et al. [30] proposed another new scheme, called *grid-group deployment scheme*. Unlike the pair-wise key scheme using deployment information in Du et al. [29], the sensors in a grid-group deployment scheme are uniformly deployed in a large area. This scheme adopts merits from the bivariate polynomial-based scheme [28] and group-based deployment scheme [29]. Similar to both schemes, a sensor deployment area of the grid-group scheme is partitioned into multiple small square areas (zones) and the sensors deployed in each zone form a group. As shown in Figure 6.3, the target deployment area is a two-dimensional rectangular deployment region with $i \times j\ a^2$ square meters. Each small deployment zone is denoted as $C(i, j)$. A sensor is identified by $[(i, j), r]$, where (i, j) is the group *id*, and r is the unique number of a sensor node.

In the key predistribution phase, based on the unconditionally secure and λ-collusion-resistant properties of the group keying scheme proposed in Blundo et al. [44], grid-group deployment schemes utilize the key predistribution scheme proposed in Du et al. [31] to distribute keys for the sensors in each zone; each node selects a sensor in each of its adjacent zones and assigns a unique key to it. After the deployment of sensors, each sensor first sets up a pair-wise key with all its neighbors within its zone

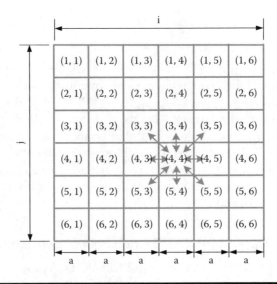

Figure 6.3   Sensor deployment in a grid structure.

by the proposed I-scheme; then it sets up a pair-wise key with its neighbors located in adjacent zones by the proposed E-scheme. Unlike R-KPS, instead of randomly distributing keys from a large key pool to each sensor, the secret keys in the grid-group deployment scheme are systematically distributed to each sensor from a structured key pool. As a consequence, the grid-group deployment scheme is resilient to selective node capture and node fabrication attack, and it requires fewer keys preinstalled for each sensor.

6.5.4 Pair-Wise Key Predistribution Schemes with Structured Key Pool

Pair-wise key establishment is a fundermental security service in sensor networks. It enables sensor nodes to communicate securely with each other using cryptographic techniques. Structured key pool [31,32] is mainly used in a pair-wise key predistribution scheme because such a structure can efficiently improve the resilience to node capture attack.

6.5.4.1 Deterministic versus Nondeterministic Pair-Wise KPS

The deterministic pair-wise KPS requires a sensor node i to store unique public information P_i and private information S_i. During the key discovery phase, nodes i and j exchange public informations and compute the pair-wise key as $f(P_j, S_i)$ and $f(P_i, S_j)$ with $f(P_j, S_i) = f(P_i, S_j)$, respectively. This scheme can ensure the λ-secure property. That is, the coalition of no more than λ compromised sensors reveals nothing about the pair-wise key between any two noncompromised nodes.

Recently, to significantly enhance the security of deterministic approaches, some nondeterministic pair-wise KPS [31,32] have been proposed. The baisc idea of nondeterministic schemes can be viewed as the combination of the basic random KPS and the deterministic pair-wise KPS. The offline TTP randomly generates a pool of m key spaces, each of which has unique private information. Each sensor node will be assigned k out of the m key spaces. If two neighboring nodes have one or more key spaces in common, they can compute their pair-wise secret key using the corresponding deterministic scheme.

6.5.4.2 Deterministic Pair-Wise KPS: Blom's Scheme

Blom [25] proposed a key predistribution method that allows any pair of nodes in a network to find a pair-wise secret key. The mechanism of Blom's λ-secure KPS can be briefly described as follows.

During the predistribution phase, the base station first generates a $(\lambda + 1) \times N$ matrix G over a finite field $GF(q)$, where N is the size of the network. The content of G is public to any sensor node. Then the base station creates

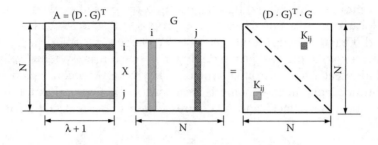

Figure 6.4   Generating pair-wise key in Blom's scheme.

a random $(\lambda + 1) \times (\lambda + 1)$ symmetric matrix D over $GF(q)$, and computes a $N \times (\lambda + 1)$ matrix $A = (D \cdot G)^T$, where $(D \cdot G)^T$ is the transpose of $D \cdot G$. Matrix D needs to be kept secretly. Because D is symmetric, the following derivations hold:

$$A \cdot G = (D \cdot G)^T \cdot G = G^T \cdot D^T \cdot G = G^T \cdot D \cdot G = (A \cdot G)^T$$

This indicates that $K = A \cdot G$ is a symmetric matrix with $K_{ij} = K_{ji}$, where k_{ij} is the element in K located in the i-th row and j-th column. Then, K_{ij} (or K_{ji}) is the pair-wise key between nodes i and j. Figure 6.4 illustrates how the pair-wise key is derived. This can be achieved using the following key predistribution approach: for $k = 1, 2, \ldots, N$, node k stores the k-th row of matrix A and the k-th column of matrix G, respectively.

Once nodes i and j need to find the pair-wise key between them, they first exchange their columns of G. So they can use their private rows of A to compute k_{ij} and k_{ji} with $K_{ij} = K_{ji}$, respectively. Because G is public, its column can be transmitted in plaintext. It has been proved [4] that the above scheme is λ-secure if any $\lambda + 1$ columns of G are linearly independent. This λ-secure property guarantees that all communication links of noncompromised nodes remain secure if no more than λ nodes are compromised.

6.5.4.3   *Deterministic Pair-Wise KPS: Blundo's Scheme*

Blundo's scheme [44] is a polynomial-based KPS, which is initially developed for group key predistribution. The key setup server randomly generates a bivariate t-degree polynomial $f(x, y) = \sum_{i,j=0}^{t} a_{ij} \cdot x^i \cdot y^j$ over a finite field F_q, where q is a prime number that is large enough to accommodate a cryptographic key. The polynomial has the symmetric property such that $f(x, y) = f(y, x)$. It is assumed that each sensor has a unique ID. For sensor i, the setup server computes a *polynomial share* of $f(x, y)$ (i.e., $f(i, y)$). Then, for any two nodes i and j, node i can obtain the common key $f(i, j)$ by evaluating at point j, and node j can compute the same key $f(j, i)$ by evaluating $f(j, y)$ at point i.

This approach requires each sensor node i to store a t-degree polynomial $f(i, x)$ that occupies $(t + 1) \log q$ storage spaces. A nice property of this scheme is that there is no communication overhead during the process of establishing a pair-wise key, because both sensor nodes could evaluate the polynomial using the ID of the other sensor node.

According to the security proof in Blundo's scheme, this scheme is unconditionally secure and t-collusion resistant. That is, if the coalition of the compromised sensor nodes is no more than t, the pair-wise key between any two noncompromised nodes will not be revealed.

Although the group key distribution scheme of Blundo et al. [44] is theoretically possible to use in WSNs, the storage cost for a polynomial share is exponentially increased with the group size, making it prohibitive in sensors with low memory capacity. Some improved solutions must be introduced.

6.5.4.4 Nondeterministic Pair-Wise KPS: Du et al.'s Scheme

To achieve better resilience against node capture, Du et al. [31] proposed a nondeterministic pair-wise KPS to significantly enhance the security of Blom's deterministic approaches. The idea of introducing multiple key spaces can be viewed as the combination of the basic random key predistribution scheme and Blom's deterministic approaches.

Du et al.'s scheme consists of two phases: (1) *key predistribution phase* and (2) *key agreement phase*. The *key predistribution* phase contains the following steps:

1. *Generate matrix G:* The scheme first creates a generator matrix G of size $(\lambda + 1) \times N$, and node j will be assigned with $G(j)$, the j-th column of matrix G.
2. *Generate matrix D:* This step generates ω symmetric matrices D_1, $D_2, \ldots, D_\omega$ of size $(\lambda + 1) \times (\lambda + 1)$. Each tuple, $S_i = (D_i, G)$, $i = 1, \ldots, \omega$, forms a key space, and $A_i = (D_i \cdot G)^T$.
3. *Choose τ key space:* τ distinct key spaces are randomly selected from the ω key spaces for each node. For each space S_i selected by node j, the j-th row of A_i (i.e., $A_i(j)$) will be stored at this node. According to Blom's scheme, two nodes can find a common secret key if they both picked a common key space.

In the *key agreement* phase, each node needs to discover whether it shares any key space with its neighbors. To do so, each node broadcasts a message containing the node's ID, the indices of the space it carries, and the seed of the column of G it carries. Assume that nodes i and j are neighbors. If they find out that they have a common key space (e.g., S_c), they can compute their pair-wise key using Blom's scheme: initially node i has $A_c(i)$ and seed for $G(i)$ and node j has $A_c(j)$ and seed for $G(j)$. After exchanging the seeds, node i and node j regenerate $G(j)$ and $G(i)$, respectively. Then the pair-wise secret key between nodes i and j, can be computed as follows:

$$K_{ij} = K_{ji} = A_c(i) \cdot G(j) = A_c(j) \cdot G(i)$$

Once the pair-wise keys with the neighbors are set up, the entire network forms a *key-sharing graph* $G_{ks}(V, E)$, where V is a set of nodes and E is a set of edges. The key graph is constructed as follows: for any two nodes i and j in V, there exists an edge between them iff (1) nodes i and j have at least one common key space, and (2) nodes i and j can reach each other within the signal range.

In Du et al.'s scheme, any two neighboring nodes i and j without sharing a common key space could still establish a pair-wise key between them. The idea is to use the secure links that are already created in the key-sharing graph $G_{ks}(V, E)$. Because $G_{ks}(V, E)$ is connected, any two neighboring nodes, say a and b, can always find a direct or indirect path $(a, v_1, \ldots, v_t, b)$ in $G_{ks}(V, E)$. To establish a common secret key between them, node a first generates a random key K. Then a sends K to v_1 using the secure link (a, v_1); v_1 also relays K to v_2 using (v_1, v_2), and so on, until b receives K from v_t. Finally, nodes a and b use K as their pair-wise key.

To make it possible for any pair of nodes to find a secret key between them, the graph $G_{ks}(V, E)$ must be connected. Given the size and the density of a network, to guarantee that $G_{ks}(V, E)$ is connected with high probability, the parameters ω and τ should be carefully selected according to the following inequation:

$$1 - \frac{[(\omega - \tau)!]^2}{(\omega - 2\tau)!\omega!} \geq \frac{N - 1}{n \cdot N}[\ln N - \ln(-\ln P_c)]$$

where N is the size of the network and P_c is the given global connectivity. In addition, the actual local connectivity P_{actual} is also determined by ω and τ; that is:

$$P_{actual} = 1 - \frac{[(\omega - \tau)!]^2}{(\omega - 2\tau)!\omega!}.$$

The collection of space sets assigned to each sensor form a probabilistic quorum system. To achieve a desired P_{actual} for a given ω, τ must satisfy

$$\tau \geq \sqrt{\ln \frac{1}{1 - P_{actual}}} \cdot \sqrt{\omega}$$

6.5.4.5 Nondeterministic Pair-Wise KPS: Liu and Ning's Scheme

The polynomial-based KPS discussed in the above section has some limitations. In particular, it can tolerate no more than t compromised nodes, where the value of t is limited by the memory available in sensor nodes. Indeed, the larger a sensor network is, the more likely an adversary compromises more than t sensor nodes and then the entire network.

Liu et al. [32] developed a general framework for key predistribution based on Blundo's scheme. This framework is called *polynomial pool-based key predistribution*, because a pool of multiple random bivariate polynomials are used in this framework. The basic idea of polynomial pool-based KPS can be considered as the combination of polynomial-based key predistribution and the key pool idea used in [4,27]. Intuitively, this general framework uses a pool of randomly generated bivariate polynomials to help establish pair-wise keys between sensors. The polynomial pool has two special cases. If the polynomial pool has only one polynomial, the general framework will degenerate into the polynomial-based KPS. If all the polynomials are 0-degree ones, the polynomial pool will degenerate into a random key pool [4,27].

Pair-wise key establishment in this scheme is performed in three phases: (1) *setup*, (2) *direct key establishment*, and (3) *path key establishment*. The *setup phase* initializes the sensors by distributing polynomial shares to them. After being deployed, if two sensors need to establish a pair-wise key, they first attempt to do so through *direct key establishment*. If they can successfully establish a common key, there is no need to start *path key establishment*. Otherwise, these sensors start path key establishment, trying to establish a pair-wise key with the help of the other sensors.

Based on the basic framework, Liu et al. proposed two instantiations of the general framework: (1) *random subset assignment* and (2) *grid-based key predistribution* scheme. The *random subset assignment KPS* can be considered an extension to the basic probabilistic scheme in Chan et al. [4]. Instead of randomly selecting keys from a large key pool and assigning them to sensors, the random subset assignment KPS randomly chooses polynomials from a polynomial pool and distributes their polynomial shares to sensors.

Assume that in the setup phase, the setup server randomly generates a set F of s bivariate t-degree polynomials over the finite field F_q. For each sensor, the setup server randomly picks a subset s' of polynomials from F and assigns polynomial shares of these s' polynomials to the sensor nodes. Similar to the analysis in Eschenauer and Gligor [27], the probability of two sensors sharing the same bivariate polynomials, which is the probability that two sensors can establish a pair-wise key directly, can be estimated by:

$$p = 1 - \prod_{i=0}^{s'-1} \frac{s - s' - i}{s - i}$$

The probability of two sensor nodes establishing a pair-wise key through direct polynomial share discovery or indirect path-key discovery can be calculated as:

$$P_s = 1 - (1 - p)(1 - p^2)^d$$

where d denotes the average number of neighboring nodes that each sensor node can contact.

According to the security analysis in Blundo et al. [44], an attacker cannot determine noncompromised keys if no more than t sensor nodes have been compromised. If an attacker randomly compromises N_c sensors, where $N_c \geq t + 1$, then the probability of any polynomial being compromised is:

$$P_c = 1 - \sum_{i=0}^{t} \frac{N_c!}{(N_c - i)!i!} \left(\frac{s'}{s}\right)^i \left(1 - \frac{s'}{s}\right)^{N_c - i}$$

where $\frac{s'}{s}$ denotes the probability of any polynomial f in being chosen for a sensor node. Because f is any polynomial in F, the fraction of compromised links between noncompromised sensors can be estimated as P_c.

Compared to the random subset assignment KPS, *grid-based KPS* offers a number of attractive properties:

1. It guarantees that any two sensor nodes can establish a pair-wise key when there are no compromised sensors, provided that the sensor nodes can communicate with each other.
2. It is resilient to node compromise. Even if some nodes are compromised, it is highly probable to establish a pair-wise key between noncompromised sensors.
3. A sensor can directly determine whether it can establish a pair-wise key with another node. As a result, there is no communication overhead during polynomial share discovery. If a senor network has at most N nodes, the three phases of grid-based KPS can be described as follows.

In *subset assignment phase* of the grid-based key KPS, it constructs a $m \times m$ grid with a set of $2m$ polynomials $\{f_i^r(x, y), f_j^c(x, y)\}$, $i, j = 0, 1, \ldots, m - 1$, where $m = \lceil \sqrt{N} \rceil$. Each row and each column in the grid are associated with a polynomial, respectively. Each sensor is assigned to a unique intersection in this grid. Accordingly, the polynomial shares are distributed to the sensor at the intersection coordinate. Based on this information, each sensor node thus can perform share discovery and path discovery.

The goal of the *polynomial share discovery* phase is to establish a pair-wise key. For node j, node i checks if $c_i = c_j$ or $r_i = r_j$. If $c_i = c_j$, they can use the polynomial-based KPS to establish a pair-wise key directly, because both node i and j have common polynomial shares of $f_{c_j}^c(x, y)$. Similarly, if $r_i = r_j$, this means that they both have polynomial shares of $f_{r_j}^r(x, y)$ and can also establish a pair-wise key. If neither of these conditions is true ($c_i \neq c_j$ and $r_i \neq r_j$), node i and j can only indirectly establish a pair-wise key during the *path discovery phase*.

6.5.4.6 Combinatorial-Based R-KPS

Combinatorial-based KPS is based on two concepts: (1) *generalized quadrangle* and (2) *symmetric design* [45]. A *generalized quadrangle* is an incidence structure $S = (P, B, I)$, where P and B are disjoint nonempty sets of points and lines respectively, and I is a symmetric point-line incidence relation satisfying the following axioms:

1. Each point is incident with lines and two distinct points are incident with at most one line.
2. Each line is incident with $t + 1$ points and two distinct lines are incident with at most one point.
3. If x is a point and L is a line not incident (I) with x, then there is a unique pair $(y, M) \in P \times B$, for which $x I M I y I L$.

Symmetric design is a special case of *balanced incomplete block design* (BIBD). The BIBD is an arrangement of v distinct objects into b blocks such that each block contains exactly k distinct objects, each object occurs in exactly different blocks, and every pair of distinct objects occurs together in exactly λ blocks. The design can be expressed as (v, k, λ), or equivalently (v, b, r, k, λ), where $\lambda(v - 1) = r(k - 1)$ and $bk = vr$. The BIBD is called *symmetric design* when $b = v$ and thus $r = k$.

According to the two basic combinatorial block designs, Qamtepe and Yener [33] first introduce two algorithms to map the two classes of combinatorial designs, *symmetric design* and *generalized quadrangle*, to the deterministic key distribution mechanism. The main drawback of the combinatorial approach comes from the difficulty in their construction. Given a desired number of sensor nodes or a desired number of keys in the pool, it may not be able to construct a combinatorial design for the target parameters.

To address the issues, based on the two basic mapping algorithms, Qamtepe and Yener further present two *hybrid design* approaches, *hybrid symmetric design* and *hybrid GQ design*, by combining the deterministic core and probabilistic extension. These two novel schemes preserve the positive properties of combinatorial design, yet take advantage of the flexibility and scalability of probabilistic approaches to support the key distribution in any network size.

The performance comparisons of different designs can be summarized as follows. For the same network size, key chain size, and pool size, *symmetric combinatorial design* provides a better probability of key sharing between any two key chains. In R-KPS, a pair of nodes is required to go through a path of 1.35 hops, on average, to share a key and communicate securely, while the key path length is only one for the symmetric combinatorial method. Compared with R-KPS, the *generalized quadrangle* scheme decreases key chain size, causing a small decrease in key sharing probability. Although R-KPS provides slightly better probability of key sharing between any two key chains, generalized quadrangle is still competitive with R-KPS. Because when two chains do not share a key, it can guarantee the existence of a third one that shares a key with both. A *hybrid symmetric combinatorial* scheme makes use of the symmetric combinatorial method, yet introducing the scalability of the probabilistic approach. Hence, given a target network size and key chain size, a hybrid symmetric combinatorial scheme shows better performance than R-KPS.

As a whole, Qamtepe and Yener [29] argue that combinatorial-based KPS has the following advantages: (1) increased probability of a pair of sensors sharing a key, and (2) decreased key-path length, which provides scalability with hybrid approaches. Thus, these approaches can offer better connectivity with smaller key chain sizes.

6.5.5 Path-Key Establishment Protocol

Chan et al. [4] proposed a *multi-path key reinforcement scheme*. It can substantially enhance the security of a pair-wise key by establishing the path key through multiple paths, because an attacker has to compromise many more nodes to achieve a high probability of compromising any given communication. This method can also be applied in conjunction with the basic random KPS to yield greatly improved resilience against node capture attacks by trading off some network communication overhead.

The main advantage of a multi-path key reinforcement scheme is simple. Suppose source node u requires establishing a secure link to destination node v after key setup. To resist a node capture attack, multiple physically link-disjoint paths between two nodes are used to set up the path key. Suppose that a and b intend to set up a pair-wise key k via multiple (say $j > 1$) link-disjoint paths. The source node a will randomly select j secrets $s_1, s_2, \ldots, s_j$ with $k = s_1 \oplus s_2 \oplus \ldots \oplus s_j$, and sends each secret to b via a unique key establishment path. For example, for a given secret s_1 transferred via a path $a \rightarrow x \rightarrow b$, the following steps are executed: $a \rightarrow x$: $\{s_1\}_{k_{ax}}; x \rightarrow b$: $\{s_1\}_{k_{xb}}$, where k_{ax} and k_{xb} are pair-wise keys shared between pair (a, x) and (x, b), respectively. Once node b has received all j secrets, node b can simply derive the pair-wise key as $k = s_1 \oplus s_2 \oplus \ldots \oplus s_j$. Thus, the secrecy of the pair-wise key k is protected by all j random secrets. Unless the attack can successfully eavesdrop on all j paths, it will not steal sufficient pieces of the pair-wise key to recover it.

Another multi-path scheme is given in Zhu et al. [34], in which a pair-wise key is set up using multiple logical link-disjoint paths between these two nodes. A logical path denotes that there are shared key relations among source, destination, and intermediate nodes along the key establishment path. There exist two types of logical paths between two nodes: (1) *direct path*, where two nodes share one or more keys in their key ring, and (2) *indirect path*, where two nodes do not share any key, but through other intermediate nodes they can exchange messages securely. Suppose that source node a shares t_1 direct keys with intermediate node x, and node x shares t_2 direct keys with the destination node b (nodes a and b do not share a direct key). Then there are only $n = \min(t_1, t_2)$ key paths between nodes a and b via node x, because a direct key can be only used for one logical path.

The secret generation approach in Zhu et al. [34] is similar to that in Chan et al. [4]. The main differences include:

1. The use of physical or logical key establishment paths in respective schemes.
2. In Zhu et al. [34], an efficient algorithm is presented to determine the reasonable number of secret shares to be used for establishing a pair-wise key according to the required security level.
3. In Zhu et al. [34], during the key predistribution phase, the set of keys assigned to each sensor is computed using a pseudo-random generator with the sensor ID as the seed. Such an ID-based de-terministic key assignment algorithm allows any sensor to know the set of keys stored by another sensor. This feature supports a key discovery phase that involves only local computations with-out message exchange. Thus, it is a more communication-efficient scheme.

Although both of the above proposed multi-path key establishment schemes are efficient in resisting outsiders' node capture attacks and passive Byzantine attacks [10], they are still vulnerable to active Byzantine attacks. An attack can prevent the destination node from deriving the right pair-wise key by stopping the forwarding of secrets or altering the forwarding se-crets. To address this issue, Huang et al. [35] proposed a Byzantine-resilient multi-path key establishment scheme using the Reed-Solomon error-correct coding scheme. This scheme offers the following advantages:

1. It can tolerate, at most, $t = (n - k)/2$ faulty key paths, when the (n, k) Reed-Solomon error-correct coding scheme is used.
2. It is a communication-efficient scheme. The receiving sensors can identify the faulty paths with minimal communication overhead, because no interactive communications are required to identify such faulty key establishment paths.

The robustness of Huang et al.'s scheme [36] is defined as the probability that a pair-wise key is not compromised if there are x compromised nodes. Huang et al. analyze two styles of attacks. One is the resilience against passive key establishment attacks. Suppose that there are x compromised nodes that allude by sharing their key sets. Then, the probability that at least one of the direct keys on an h hop path is compromised is estimated as $p_{passive}(x, h) = 1 - (1 - m/P)^{xh}$. Accordingly, if all key paths have the same length h, the robustness against passive key establishment attacks can be computed as:

$$R_{passive} = \sum_{i=0}^{t} \binom{p}{i} [p_{passive}(x, h)]^i [1 - p_{passive}(x, h)]^{p-i}.$$

The other is the resilience against active key establishment attacks. For a key path with h hops, the probability that at least one of x compromised nodes can create a short cut on a key path is:

$$p_{active}(x, h) = 1 - \left\{ 1 - \left[1 - \frac{\binom{\omega}{r}\binom{\omega-r}{r}}{\binom{\omega}{r}^2} \right]^2 \right\}^{th}$$

Accordingly, if all key paths have the same length h, the robustness against active key establishment attacks can be computed as:

$$R_{active} = \sum_{i=0}^{t} \binom{p}{i} [p_{active}(x, h)]^i [1 - p_{active}(x, h)]^{p-i}.$$

Evidently, increasing the threshold t will increase the resilience to Byzantine attacks.

6.5.6 Improved Shared-Key Discovery Schemes

As described above, all R-KPS require a key discovery phase. So far, three different schemes have been proposed to improve the security and performance: (1) key index notification, (2) challenge-response, and (3) pseudo-random key index transformation. The corresponding communication complexity of the above discovery procedures is summarized in Table 6.1.

Key index notification [27] requires sensor a to send sensor b the indexes of the keys in its key ring. Sensor a, in turn, notifies sensor b of the indexes of the keys to be used to secure the channel. The communication complexity for sensor a is equal to K messages, where K is the size of the key ring. The computational complexity for sensor b is also equal to

Table 6.1 Complexity of the Key Discovery Phase Scheme

	Sender		Receiver		
Schemes	*Msg.*	*Computations*	*Msg.*	*Computations*	*Key Ring Size*
Index notification [27]	K	K	K	$K \log(K)$	K
Challenge response [27]	K	K encryption	K	K^2	K
Pseudo-random index [34]	0	$K \log(K) +$ K hash	0	$K \log(K) +$ K hash	K
Pseudo-random index [36]	0	K hash	0	K hash	$K + O(K)$

K, because both sensor a and b store their keys sorted according to their indexes.

Unlike the key index notification method, *challenge-response*, which is introduced in Eschenauer and Gligor [27] and later used in [4,31,32], does not leak any information about the indexes of the keys in the key ring. However, it suffers from the higher energy cost due to the commnications for message exchange. Specifically, sensor a is supposed to broadcast messages, α, $E_{K_i}(\alpha)$, $i = 1, \ldots, k$, where α is a challenge and K_i is the i-th key assigned to sensor a from the pool. The decryption of $E_{K_i}(\alpha)$ with the proper key by sensor b would reveal the challenge α and the information that sensor b shares that particular key with sensor a. This key discovery procedure requires $O(k^2)$ decryptions on the receiver side and k encryptions on the sender side. Moreover, at least K messages must be sent or received.

Compared with the key index notification and challenge-response approaches, *pseudo-random key index transformation* [34,36] offers a more efficient key discovery procedure. It requires no communications in the key discovery phase and offers high resilience against the node capture. The mechanism of this approach is illustrated as follows. Given a sensor a, the indexes of the keys are derived using a pseudo-random generator (e.g., one-way hash function) initialized with a public seed dependent on a. Then all the key indexes will be assigned to a. Once the public seed is known, the k indexes of the keys assigned to a can be computed by anyone. Note that the key discovery procedure supported by pseudo-random transformation reveals only the indexes of the keys given to sensor a, and does not leak any information on the keys themselves. This key discovery procedure requires no message exchange, at most k applications of the pseudo-random generator, and k look-ups in the local memory of each consisting of $\log(k)$ operations.

Both key index notification and pseudo-random key index transformation have been criticized for revealing the key indexes of the keys in each sensor's key ring to the attacker [4,27]. It is argued, intuitively, that knowing such indexes allows the attacker to pick and tamper with exactly those sensors whose key rings are most useful for compromising a given target channel. However, neither mathematics nor experimental justifications have been given to support this intuition. According to Table 6.1 [36], it seems that the pseudo-random KPS can assign keys to sensors more efficiently, offer an energy-preserving key discovery phase that is communication-efficient, and is more resilient against node capture attacks.

6.6   Some Open Issues

As mentioned, research in the area of key management for WSNs is far from exhaustive. Much of the effort thus far has been on devising key management protocols to support effective key distribution or agreement. However, there are still many issues that deserve further investigation.

- **Scalability**: This issue is not only related to key management in WSNs, but also to the WSN itself. Can a key management protocol be designed for WSNs that is scalable with respect to the number of sensor nodes, their mobility, and other constraints posed by the WSN environment itself?
- **Key revocation or dynamic update**: A new sensor should be dynamically deployed in a WSN, and a detected misbehaving node should also be dynamically removed from the system. An obvious question is how to efficiently revoke or update a expire key between sensor nodes.
- **Fault tolerance**: The dynamics of WSNs and the easy-to-impair communication environment also increase service interruption probability, which demands that the scheme is robust against packet loss. Several related works can be found in References [46–48].
- **DoS attack tolerance**: The unattended deployment environment and the limited computation capacity in sensor nodes make them vulnerable to denial-of-service (DoS) attacks [6,7], which can diminish or eliminate network capacity to perform its expected function. How to provide a DoS tolerance architecture deserves careful investigation.
- **No TTP KPS**: All existing KPS schemes still rely on TTP (online or offline) in the initialization phase to pre-load each node with the long-lived keys and the corresponding key identifiers. Clearly, relying on TTP compromises the security of the whole network. This limitation renders those schemes inapplicable in many ad hoc

scenarios although the TTP is only needed during initialization and can be offline the rest of the time. Naturally, how to introduce practical schemes to realize fully distributed and self-organized KPS is a challenge. Chan [49] makes the primary attempt for distributed key predistribution without relying on TTPs and establishes a routing infrastructure, but Wu et al. [50] point out that this scheme is falsely deduced.

■ **Public key-based management schemes**: The most common criticism of using public key cryptography (PKC) in sensor networks is its computational complexity and communication overhead. However, with the advance of technology, PKC may sooner or later be widely used in WSNs. Recently, a number of studies have shown that the performance of some public key algorithm, such as elliptic curve cryptography [51], is already close to being practical for sensor nodes [52–55]. However, the energy consumption of PKC is still expensive, especially when compared to symmetric cryptography. To maximize the lifetime of batteries, it is required to minimize the use of PKC whenever possible in WSNs. How to reduce the amount of expensive public-key operations in PKC-based schemes for sensor networks may be a promising research area. Some schemes proposed for traditional ad hoc networks possibly can be used in WSNs [56–59].

6.7 Conclusions

Recent advances in wireless communications have mobilized the development of economic sensor networks for various applications. The flexibility, fault tolerance, high sensing fidelity, low cost, and rapid development characteristics of sensor networks create many new and exciting application areas for remote sensing. Efficient key management, as one of the fundamental building blocks, plays a vital role in security services. However, to implement an efficient key management scheme for WSNs, we need to satisfy the constraints introduced by factors such as scalability, fault tolerance, low computation capacity, topology change, and power consumption. Because these constraints are highly stringent and specific for sensor networks, the existing key management techniques are still not enough to implement efficient key distribution, and some new efficient key management schemes are urgently required. In this chapter we have provided a thorough description of the current state-of-the-art in key management schemes for WSNs. The usefulness of different protocols depends on the application environment. Much work must be done before we have consistent and adequate solutions to key management in WSNs. Issues, such as fully distributed, self-organized, energy-aware, and location-aware key management schemes, need to be investigated in much more detail. On the other hand, with the advances in technology, the low resource constraints of sensor nodes may not be a concern. So, there is high potential that a key distribution scheme and a key agreement scheme will be adopted in WSNs in the future.

References

1. I.F. Akyildiz, W. Su, Y. Sankarasubramaniam, and E. Cayirci, A Survey on Sensor Networks, *IEEE Commun. Magazine*, Vol. 40, No. 8, pp. 102–114, 2002.
2. S. Basagni, K. Herrin, D. Bruschi, and E. Rosti, Secure Pebblenets, *Proc. of the 2nd ACM International Symposium on Mobile Ad Hoc Networking & Computing*, pp. 156–163, 2001.
3. J. Kahn, R. Katz, and K. Pister, Next Century Challenges: Mobile Networking for Smart Dust, *Proc. of ACM MobiCom99*, pp. 483–492, 1999.
4. H. Chan, A. Perrig, and D. Song, Random Key Pre-distribution Schemes for Sensor Networks, *Proc. of IEEE Symposium on Security and Privacy*, pp. 197–213, 2003.
5. D.W. Carman, P.S. Kruus, and B.J. Matt, Constraints and Approaches for Distributed Sensor Security, *NAI Labs Technical Report*, #00-010, 2000.
6. A.D. Wood and J.A. Stankovic, Denial of Service in Sensor Networks, *Computer*, Vol. 35, No. 10, pp. 54–62, October 2002.
7. M. Long, C.H. Wu, and J.Y. Hung, Denial of Service Attacks on Network-Based Control Systems: Impact and Mitigation, *IEEE Trans. on Industrial Informatics*, Vol. 1, No. 2, pp. 85–96, May 2005.
8. W. Diffie and M.E. Hellman, New Directions in Cryptography, *IEEE Trans. on Information Theory*, IT-22, pp. 644–654, 1976.
9. R.L. Rivest, A. Shamir, and L.M, Adleman, A Method for Obtaining Digital Signatures and Public-Key Cryptosystems, *Communications of the ACM*, Vol. 21, No. 2, pp. 120–126, 1978.
10. L. Lamport, R. Shostak, and M. Pease, The Byzantine Generals Problem, *ACM Trans. on Programming Languages and Systems*, Vol. 4, No. 3, pp. 382–401, 1982.
11. J. Kohl and B. Neuman, The Kerberos Network Authentication Service (V5), IETF RFC 1510, 1993.
12. B.C. Neuman and T. Tso, Kerberos: An Authentication Service for Computer Networks, *IEEE Commun.*, Vol. 32, No. 9, pp. 33–38, 1994.
13. B. DeCleene, L. Dondeti, et al., Secure Group Communications for Wireless Networks, *Proc. of IEEE MILCOM01*, 2001.
14. S.P. Griffin, B.T. DeCleene, L.R. Dondeti, R.M. Flynn, D. Kiwior, and A. Olbert, Hierarchical Key Management for Mobile Multicast Members, Technical Report, Northrop Grumman Information Technology, 2002.
15. M. Striki and J.S. Baras, Key Distribution Protocols for Multicast Group Communication in MANETS, Technical Report #TR2003-17, University of Maryland, 2003.

16. G. Ateniese, M. Steiner, and G. Tsudik, New Multiparty Authentication Services and Key Agreement Protocols, *IEEE J. on Select. Areas in Commun.*, Vol. 18, No. 4, pp. 628–640, 2000.

17. K. Becker and U. Wille, Communication Complexity of Group Key Distribution, *Proc. of ACM CCS98*, pp. 1–6, 1998.

18. M. Burmester and Y. Desmedt, A Secure and Efficient Conference Key Distribution System, *Proc. of Eurocrypt94*, pp. 275–286, 1994.

19. R. Canetti, J. Garay, G. Itkis, D. Micciancio, M. Naor, and B. Pinkas, Multicast Security: A Taxonomy and Some Efficient Constructions, *Proc. of INFOCOM99*, pp. 708–716, 1999.

20. I. Ingemarsson, D.T. Tang, and C.K. Wong, A Conference Key Distribution System, *IEEE Trans. on IT*, Vol. 28, No. 5, pp. 714–720, 1982.

21. Y. Kim, A. Perrig, and G. Tsudik, Communication-Efficient Group Key Agreement, *Proc. of the 16th International Conf. on Information Security: Trusted Information: The New Decade Challenge, IFIP SEC 2001*, Paris, France, pp. 229–244, 2001.

22. M.S. Hwang, Dynamic Participation in a Secure Conference Scheme for Mobile Communication, *IEEE Trans. on Vehicular Tech.*, Vol. 48, No. 5, pp. 1469–1474, 1999.

23. K.F. Hwang and C.C. Chang, A Self-Encryption Mechanism for Authentication of Roaming and Teleconference Services, *IEEE Trans. on Wireless Commun.*, Vol. 2, No. 2, pp. 400–407, 2003.

24. X. Yi, C.K. Siew, C.H. Tan, and Y. Ye, A Secure Conference Scheme for Mobile Communication, *IEEE Trans. on Wireless Commun.*, Vol. 2, No. 6, pp. 1168–1177, 2003.

25. R. Blom, An Optimal Class of Symmetric Key Generation Systems, *Advances in Cryptology: Proc. of EUROCRYPT84*, LNCS, Vol. 209, pp. 335–338, 1984.

26. T. Matsumoto and H. Imai, On the Key Predistribution Systems: A Practical Solution to the Key Distribution Problem, *Proc. of Advances in Cryptology Crypto 87*, LNCS, Vol. 293, pp. 185–193, 1988.

27. L. Eschenauer and V.D. Gligor, A Key-Management Scheme for Distributed Sensor Networks, *Proc. of ACM CCS02*, pp. 41–47, 2002.

28. D. Liu and P. Ning, Location-based Pair-wise Key Establishments for Static Sensor Networks, *Proc. of the 1st ACM Workshop on Security of Ad Hoc and Sensor Networks*, Virginia, pp. 72–82, 2003.

29. W. Du, J. Deng, Y.S. Han, S. Chen, and P.K. Varshney, A Key Management Scheme for Wireless Sensor Networks Using Deployment Knowledge, *Proc. of IEEE INFOCOM04*, Vol. 1, pp. 586–597, 2004.

30. D. Huang, M. Mehta, D. Medhi, and L. Harn, Location-Aware Key Management Scheme for Wireless Sensor Networks, *Proc. of the 2nd ACM Workshop on Security of Ad Hoc and Sensor Networks*, October 25, 2004, Washington, D.C., pp. 29–42, 2004.

31. W. Du, J. Deng, Y.S. Han, and P.K. Varshney, A Pair-wise Key Predistribution Scheme for Wireless Sensor Networks, *Proc. of ACM CCS03*, pp. 42–51, 2003.

32. D. Liu and P. Ning, Establishing Pair-wise Keys in Distributed Sensor Networks, *Proc. of ACM CCS03*, pp. 52–61, 2003.

33. S.A. Camtepe and B. Yener, Combinatorial Design of Key Distribution Mechanisms for Wireless Sensor Networks, *Proc. of 9th European Symp. on Research in Computer Security (ESORICS'04)*, pp. 293–308, 2004.

34. S. Zhu, S. Xu, S. Setia, and S. Jajodia, Establishing Pair-wise Keys for Secure Communication in Ad Hoc Networks: A Probabilistic Approach, *Proc. of the 11th IEEE International Conf. on Network Protocols*, pp. 326–335, November 04–07, 2003.

35. D. Huang and D. Medhi, Byzantine Resilient Multi-path Key Establishment in Sensor Networks, *Proc. of 18th International Parallel and Distributed Processing Symp.* (IPDPS05), pp. 240–247, 2005.

36. R.D. Pietro, L.V. Mancini, and A. Mei, Efficient and Resilient Key Discovery Based on Pseudo-random Key Pre-deployment, *Proc. of 18th International Parallel and Distributed Processing Symp. (IPDPS04)*, pp. 217–214, 2004.

37. M. Mehta, D. Huang, and L. Harn, A Practical Scheme for Random Key Pre-distribution Scheme and Shared-Key Discovery in Sensor Networks, *Proc. of 24th IEEE International Conf. on Performance Computing and Commun.*, 2004.

38. P. Erds and A. Rnyi, On Random Graphs I, *Publ. Math. Debrecen*, pp. 290–297, 1959.

39. J. Spencer, The Strange Logic of Random Graphs, *Algorithms and Combinatorics 22*, Springer-Verlag, 2000.

40. R.D. Pietro, L.V. Mancini, A. Mei, A. Panconesi, and J. Radhakrishnan, Connectivity Properties of Secure Wireless Sensor Networks, *Proc. of the 2nd ACM Workshop on Security of Ad Hoc and Sensor Networks*, Washington, D.C., pp. 53–58, 2004.

41. J. Hwang and Y. Kim, Revisiting Random Key Pre-distribution Schemes for Wireless Sensor Networks, *Proc. of the 2nd ACM Workshop on Security of Ad Hoc and Sensor Networks*, pp. 43–52, 2004.

42. R.D. Pietro, L.V. Mancini, and A. Mei, Random Key-Assignment for Secure Wireless Sensor Networks, *Proc. of the 1st ACM Workshop on Security of Ad Hoc and Sensor Networks*, 2003, Virginia, pp. 62–71, 2003.

43. F. Xue and P.R. Kumar, The Number of Neighbors Needed for Connectivity of Wireless Networks, *Wireless Networks*, Vol. 10, No. 2, pp. 169–181, 2004.

44. C. Blundo, A.D. Santis, A. Herzberg, S. Kutten, U. Vaccaro, and M. Yung, Perfectly-Secure Key Distribution for Dynamic Conferences, *Proc. of the 12th Annual International Cryptology Conference on Advances in Cryptology*, LNCS, Vol. 740, pp. 471–486, 1992.

45. I. Anderson, *Combinatorial Designs: Construction Methods*, Ellis Horwood Limited, 1990.

46. A. Perrig, R. Szewczyk, V. Wen, D. Culler, and J. D. Tygar, SPINS: Security Protocols for Sensor Netowrks, *Proc. of IEEE/ACM MobiCom01*, pp. 189–199, July 2001, Italy.

47. T. Park and K.G. Shin, LiSP: A Lightweight Security Protocol for Wireless Sensor Networks, *ACM Trans. on Embedded Computing Systems*, Vol. 3, Issue 3, pp. 634–660, August 2004.

48. D. Liu and P. Ning, Multilevel TESLA: Broadcast Authentication for Distributed Sensor Networks, *ACM Trans. on Embedded Computing Systems*, Vol. 3, Issue 4, pp. 800–836, November 2004.

49. C.F. Chan, Distributed Symmetric Key. Management for Mobile Ad Hoc Networks, *Proc. of IEEE INFOCOM04*, Vol. 4, pp. 2414–2424, 2004.

50. J. Wu and R. Wei, Comments on Distributed Symmetric Key Management for Mobile Ad Hoc Networks, http://eprint.iacr.org/2005/008.pdf.

51. N. Kobliz, Elliptic Curve Cryptography, *Mathematics of Computation*, Vol. 48, 1987.

52. P. Ganesan, R. Venugopalan, P. Peddabachagari, A. Dean, F. Mueller, and M. Sichitiu, Analyzing and Modeling Encryption Overhead for Sensor Network Nodes, *Proc. of the 1st ACM International Workshop on Wireless Sensor Networks and Applications*, USA, pp. 151–159, 2003.

53. G. Gaubatz, J. Kaps, and B. Sunar, Public Key Cryptography in Sensor Networks — Revisited, *Proc. of the 1st European Workshop on Security in Ad Hoc and Sensor Networks (ESAS)*, 2004.

54. D.J. Malan, M. Welsh, and M. D. Smith, A Public-Key Infrastructure for Key Distribution in TinyOS Based on Elliptic Curve Cryptography, *Proc. of the First IEEE International Conference on Sensor and Ad Hoc Commun. and Networks*, California, 2004.

55. W. Du, R. Wang, and P. Ning, An Efficient Scheme for Authenticating Public Keys in Sensor Networks, *Proc. of the 6th ACM International Symp. on Mobile Ad Hoc Networking and Computing*, MobiHoc05, pp. 58–67, 2005.

56. S. Capkun, L. Buttyan, and J.P. Hubaux, Self-Organized Public-Key Management for Mobile Ad Hoc Networks, *IEEE Trans. on Mobile Computing*, Vol. 2, No. 1, pp. 52–64, 2003.

57. J.P. Hubaux, T. Gross, J.Y. Le Boudec, and M. Vetterli, Toward Self-Organized Mobile Ad Hoc Networks: The Terminodes Project, *IEEE Comm. Magazine*, January 2001.

58. L. Zhou and Z. Haas, Securing Ad Hoc Networks, *IEEE Network*, Vol. 13, No. 6, pp. 24–30, 1999.

59. J. Kong, P. Zerfos, H. Luo, S. Lu, and L. Zhang, Providing Robust and Ubiquitous Security Support for Mobile Ad Hoc Networks, *Proc. of Ninth International Conf. on Network Protocols (ICNP)*, November 2001.

Chapter 7

Key Management for Wireless Sensor Networks in Hostile Environments

Michael Chorzempa, Jung-Min Park,
Mohamed Eltoweissy, and Y. Thomas Hou

Contents

7.1   Abstract

Large-scale wireless sensor networks (WSNs) are highly vulnerable to attacks because they consist of numerous miniaturized resource-constraint devices, interact closely with the physical environment, and communicate via wireless links. These vulnerabilities are exacerbated when WSNs have to operate unattended in a hostile environment, such as battlefields. In such an environment, an adversary poses a physical threat to all the sensor nodes; that is, an adversary may capture any node compromising critical security data, including keys used for confidentiality and authentication. Consequently, it is necessary to provide key management services to WSNs in such environments that, in addition to being efficient, are highly robust against attacks. In this chapter, we illustrate a key management design for such networks by describing a self-organizing key management scheme for large-scale WSNs, called *Survivable and Efficient Clustered Keying (SECK)*. SECK is designed for managing keys in a hierarchical WSN consisting of low-end sensor nodes clustered around more capable gateway nodes. Using cluster-based administrative keys, SECK localizes the impact of attacks and considerably improves the efficiency of maintaining fresh session keys.

7.2   Introduction

Key management is crucial to the secure operation of wireless sensor networks (WSNs). A large number of keys must be managed in order to encrypt and authenticate all sensitive data. The objective of key management is to dynamically establish and maintain secure channels among communicating parties. Typically, key management solutions use administrative keys (a.k.a. key encryption keys) to securely and efficiently (re-)distribute and, at times, generate the secure channel session keys (a.k.a. data encryption keys) to the communicating parties. Session keys may be pair-wise keys used to secure a communication channel between two nodes that are in direct or indirect communication [3,4,19,21], or they may be group keys [17,18,31,32] shared by multiple nodes. Network keys (both administrative and session keys) may need to be changed (re-keyed) to maintain secrecy and resiliency to attacks, failures, or network topology changes. Key management entails the basic functions of generation, assignment, and distribution of network keys. It is to be noted that re-keying is comprised essentially of these basic functions.

The success of a key management scheme is determined, in part, by its ability to efficiently survive attacks on the highly vulnerable and resource challenged sensor networks. Key management schemes in sensor networks can be classified broadly into dynamic or static solutions based on whether re-keying (update) of administrative keys is enabled post network deployment. Schemes can also be classified into homogeneous or heterogeneous schemes with regard to the role of network nodes in the key management process. All nodes in a homogeneous scheme perform the same functionality; on the other hand, nodes in a heterogeneous scheme are assigned different roles. Homogeneous schemes generally assume a flat network model, while heterogeneous schemes are intended for both flat as well as clustered networks. Other classification criteria include whether nodes are anonymous or have predeployment identifiers and if, when (pre-, post-deployment, or both), and what deployment knowledge (location, degree of hostility, etc.) is imparted to the nodes. In this chapter we use the primary classification of static versus dynamic keying.

Recently, numerous static key management schemes have been proposed for sensor networks. Most of them are based on the seminal random key predistribution scheme introduced by Eschenauer and Gligor [3]. In this scheme, each sensor node is assigned k keys out of a large pool P of keys in the predeployment phase. Neighboring nodes may establish a secure link only if they share at least one key, which is provided with a certain probability based on the selection of k and P. A major advantage of this scheme is the exclusion of the base station in key management. However, successive node captures enable an attacker to reveal network keys and use them to attack other nodes. Subsequent extensions to that scheme include using key polynomials [21] and deployment knowledge [20] to enhance scalability and resilience to attacks.

Another emerging category of schemes, including the scheme proposed in this chapter, uses dynamic keying and employs a combinatorial formulation of the group key management problem to affect efficient re-keying [10]. While static schemes primarily assume that administrative keys will outlive the network and emphasize pair-wise session keys, dynamic schemes advocate rekeying to achieve attack resiliency in long-lived networks and emphasize group communication keys. Table 7.1 shows the primary differences between static and dynamic keying in performing key management functions. Moharram and Eltoweissy [12] provide a performance and security comparison between static and dynamic keying.

Table 7.1   Key Management Functions in Static and Dynamic Keying

(Admin. Keys Assumed)	*Static Keying*	*Dynamic Keying*
Key assignment	Once at predeployment	Multiple times
Key generation	Once at predeployment	Multiple times
Key distribution	All keys are predistributed to nodes prior to deployment	Subsets of keys are redistributed to some nodes as needed
Re-keying	Not applicable	Multiple times; requires a small number of messages
Handling node capture	Revealed keys are lost and may be used to attack other nodes	Revealed keys are altered to prevent further attacks

SECK (Survivable and Efficient Clustered Keying) is a dynamic key management scheme that is appropriate for a network with a multi-tier hierarchical architecture deployed in a hostile environment. In such a hierarchical network, the bottom tier consists of clusters of sensor nodes, each cluster consisting of many low-end nodes and a more capable cluster head node. In this chapter we focus on robust key management within a cluster. We give more details of the network architecture in Section 7.3.

In a hostile environment, a WSN operates unattended and its nodes are highly prone to capture. SECK, originally proposed in Chorzempa et al. [1], is a self-organizing scheme that sets up key associations in the network clusters, establishes pair-wise and group keys, provides efficient methods for distributing and maintaining session keys, efficiently adds and revokes nodes, provides efficient mechanisms to recover from multiple node captures, and enables location-based reclustering of nodes. Using analytical and simulation results, we show that SECK is robust against the attacks that we identify in the threat model described in Section 7.5 of this chapter. Moreover, our results show that SECK incurs low communications and storage overhead on the sensor nodes.

The most distinguishing feature of SECK is its robustness against node compromises and its ability to recover from those compromises. SECK was designed for WSNs that must be deployed in hostile environments. In such environments, we assume that an adversary poses a physical threat to all the sensor nodes; that is, an adversary may capture any node. The noteworthy advantages of SECK are twofold: (1) SECK requires low communications overhead for key distribution and maintenance; in a hostile environment, keys must be periodically updated to maintain their trustworthiness; SECK is able to efficiently reestablish keys; and (2) SECK is able to efficiently refresh keys, revoke captured nodes, and efficiently reestablish secure group communications after node captures are detected.

In Section 7.3 we describe the WSN architecture that is used throughout the chapter. We describe the fundamental design principles of SECK in Section 7.4, present the threat model in Section 7.5, and give full details of SECK in Section 7.6. In Section 7.7 we discuss SECK's robustness against node captures. In Section 7.8 we discuss energy dissipation properties of SECK. In Section 7.9 we describe related research. Finally, we summarize our findings in Section 7.10.

7.3 The Network Architecture

In a flat network, all nodes are identical and there is no predetermined architecture. Although simple and efficient for small network sizes, the flat network architecture lacks scalability. A multi-tiered architecture provides scalability, notable energy efficiency, and security benefits [6,27,28,33,34]. Recent data aggregation techniques [8], which remove redundancy in collected data, lend themselves to this hierarchical architecture. Also, WSN routing research has shown that using a multi-tiered architecture for routing can prevent premature battery depletion among nodes near the base station, because, in a flat network, these nodes receive significantly higher traffic volume than remote nodes [29,30]. A multi-tiered architecture can also improve a network's robustness against node or key captures by limiting the effects of an attack to a certain portion of the network. For example, in a multi-tiered WSN, nodes are deployed in clusters, and each cluster can establish keys independently of other clusters. Thus, a key compromise in one cluster does not affect the rest of the network. In this section we describe a two-tiered network architecture that is suitable for large-scale WSNs.

Figures 7.1a and 7.1b show the *physical* and *hierarchical* network topology for such a network, respectively. In this architecture, a small number of high-end nodes, called Aggregation and Forwarding Nodes (AFNs), are deployed together with numerous low-end sensor nodes, called micro-sensor nodes (MSNs). In addition, the network includes a globally trusted base station (BS), which is the ultimate destination for data streams from all the AFNs. The BS has powerful data processing capabilities and is directly connected to an outside network. Each AFN is equipped with a high-end embedded processor and is capable of communicating with other AFNs over long distances. The general functions of an AFN are (1) data aggregation for information flows coming from the local cluster of MSNs, and (2) forwarding the aggregated data to the next hop AFN toward the BS. An MSN is a battery-powered sensor node equipped with a low-end processor and mechanisms for short-range radio communications at low data rates. The general function of the MSN is to collect raw data and forward the information to the local AFN. The bottom tier of the network consists of multiple clusters, where each cluster is composed of numerous MSNs

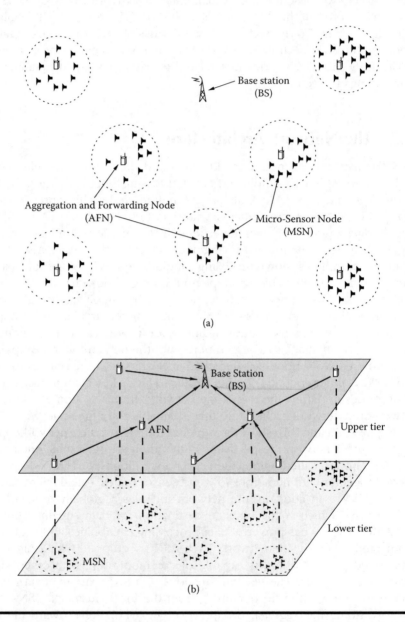

Figure 7.1 A two-tiered wireless sensor network. (a) physical topology and (b) hierarchical view.

clustered around an administrative AFN. Within each cluster, a set of keys must be deployed and managed to secure communications between the MSNs and the AFN. SECK was designed for this purpose.

7.4   Fundamental Design Principles of SECK

In this section we explain the fundamental design principles employed in SECK by describing a basic stripped-down instance of the scheme. In particular, we focus our discussions on the way SECK manages keys within a cluster. Table 7.2 presents the notation used throughout the remainder of the chapter.

To distribute and refresh session (or communication) keys, SECK assigns a set of administrative keys to each node in a network. To manage administrative keys within a cluster, SECK employs the Exclusion Basis System (EBS) [10]. EBS is based on a combinatorial formulation of the group key management problem. It essentially provides a mechanism for establishing the administrative keys held by each node. An EBS-based key management system is defined by EBS(n, k, m) where n is the number of nodes supported in the system, k is the number of keys within each key subset,

Table 7.2   Notation

Constants		Identifiers	
n	Number of nodes supported in a given EBS	N_i	i-th node
k	Keys held by each node in EBS	$K a_i$	i-th Administrative Key
m	Keys not held by each node in EBS	$K g$	Group Key
d	(Connection) degree of an AFN	$K p_i$	i-th node's Base Station Pair-wise Key
h	Number of hops between the furthest MSN and AFN_p	$K t_i$	i-th node's Tree Administrative Key
		AFN_p	Primary AFN
		AFN_b	Backup AFN

Sets		Other Notation	
U	Administrative Key set		
S_i	i-th node's Administrative Key subset		
S_{hi}	AFN's i-th hop neighbor set	$\mathbf{E}_K(M)$	Encrypt message M with key K
S_c	Compromised node set	$A \parallel B$	Bitwise concatenation of A and B
S_{ut}	Uncompromised tree set		

Table 7.3 Sample Administrative Key Subsets using EBS(10,3,2)

	N_1	N_2	N_3	N_4	N_5	N_6	N_7	N_8	N_9	N_{10}
Ka_1	1	1	1	1	1	1	0	0	0	0
Ka_2	1	1	1	0	0	0	1	1	1	0
Ka_3	1	0	0	1	1	0	1	1	0	1
Ka_4	0	1	0	1	0	1	1	0	1	1
Ka_5	0	0	1	0	1	1	0	1	1	1

and m is the number of keys from the global key set not held within a subset. We denote the administrative key set for all nodes in the system as U, where $|U| = k + m$ and the total number of keys is $k + m$. A given EBS supports $_{k+m}C_k$ unique subsets of k key assignments. For full details of the EBS, see [5,10]. We use N_i to denote the i-th node and use S_i to denote its assigned administrative key subset, with $|S_i| = k$. The notation T_i denotes the keys not held by N_i where $T_i = U - S_i$. The following property is the main motivation for utilizing the EBS. A subset of the global key set is uniquely assigned to each node such that the remaining nodes each have at least one of the keys not assigned to that node, that is, for all $j \neq i$, $S_j \cap T_i \neq \emptyset$. We will show that this property makes node revocations and session key replacement very efficient.

The AFN serving as the key management entity for its cluster must store all $k + m$ keys, and each MSN must store k keys. Note that the key subset held by each MSN is unique. This feature is utilized by the AFN to distribute session keys, that is, keys of this subset are used to encrypt the session keys before distribution. We illustrate an instance of EBS(10,3,2) in Table 7.3. The i-th ($1 \leq i \leq 10$) node in the cluster is denoted as N_i, and the j-th ($1 \leq j \leq 5$) administrative key is denoted as Ka_j. An entry marked with a "1" indicates that the node in the corresponding column possesses the administrative key of the corresponding row.

The administrative keys effectively serve as key encryption keys. They allow the AFN to establish, refresh, and revoke session keys of any *degree* belonging to any node. We define the degree of a key as the number of nodes sharing that key. An advantage of EBS is the separation of administrative keys from session keys. A node can possess any number of session keys that it may share with different subgroups of nodes. As will be described in Section 7.6, SECK uses this feature of EBS and assigns subgroup session keys in a manner that enables recovery of nodes even if all administrative keys are compromised due to the capture of multiple nodes.

As stated earlier, to initially distribute a session key to a specific node, the AFN sends that key encrypted with the unique key combination that the node possesses. To initially distribute a cluster-wide key to all members of the cluster, one message is broadcast by the AFN to all members of the cluster for each administrative key. This requires $k + m$ short broadcasts by the AFN. At any time, to update a session key Kg with Kg' and distribute Kg' to any number of members less than or equal to n, the basic version of SECK executes the aforementioned procedure using a maximum of $k + m$ keys. An example showing the update procedure of the session key, Kg, is shown below:

$$AFN \Rightarrow N_1, \ldots, N_{10} : \qquad \mathbf{E}_{Ka_1}(\mathbf{E}_{Kg}(Kg' \parallel ID_{AFN}))$$

$$AFN \Rightarrow N_1, \ldots, N_{10} : \qquad \mathbf{E}_{Ka_2}(\mathbf{E}_{Kg}(Kg' \parallel ID_{AFN}))$$

$$AFN \Rightarrow N_1, \ldots, N_{10} : \qquad \mathbf{E}_{Ka_3}(\mathbf{E}_{Kg}(Kg' \parallel ID_{AFN}))$$

$$AFN \Rightarrow N_1, \ldots, N_{10} : \qquad \mathbf{E}_{Ka_4}(\mathbf{E}_{Kg}(Kg' \parallel ID_{AFN}))$$

$$AFN \Rightarrow N_1, \ldots, N_{10} : \qquad \mathbf{E}_{Ka_5}(\mathbf{E}_{Kg}(Kg' \parallel ID_{AFN}))$$

Here, $\Rightarrow$ denotes broadcast transmission. These broadcast messages ensure that only previous holders of Kg will be able to successfully receive the new key, Kg'. If N_1 is captured, or if its keys are compromised, it is necessary to revoke all administrative and session keys held by N_1 and thus evict it from all future secure communications. This ensures forward secrecy. To accomplish this, the AFN needs to replace the administrative keys and the session keys known to N_1. From our running example, evicting N_1 would require the following two transmissions ($m = 2$).

$$AFN \Rightarrow N_1, \ldots, N_{10} :$$

$$ID_{AFN} \parallel \mathbf{E}_{Ka_1}(\mathbf{E}_{Ka_1}(Kd_1') \parallel \mathbf{E}_{Ka_2}(Kd_2') \parallel \mathbf{E}_{Ka_3}(Kd_3')),$$

$$AFN \Rightarrow N_1, \ldots, N_{10} :$$

$$ID_{AFN} \parallel \mathbf{E}_{Ka_5}(\mathbf{E}_{Ka_1}(Kd_1') \parallel \mathbf{E}_{Ka_2}(Kd_2') \parallel \mathbf{E}_{Ka_3}(Kd_3'))$$

From Table 7.3, it can be seen that all remaining nodes will be able to decipher at least one of these messages.

If more than one node is captured within a cluster, two cases arise: (1) non-colluding node captures (e.g., attacks carried out by different adversaries); and (2) colluding node captures. In the latter case, colluding attackers may compromise all administrative keys by capturing only a few nodes (e.g., by capturing nodes N_1 and N_6 shown in Table 7.3, all administrative keys are revealed). In the former case, a maximum of m^y broadcast messages will be needed to evict y nodes at once. From our running example, to evict non-colluding nodes N_1 and N_6, four messages are needed to distribute the five new keys. One message will be doubly encrypted with Ka_2 and Ka_4, the second message with Ka_2 and Ka_5, the third message with K_{a_3} and K_{a_4}, while the fourth message will be encrypted with K_{a_3} and K_{a_5}.

The basic scheme described above is efficient and functions well, provided that node captures are non-colluding. But in practice, collusion attacks can occur and the key management solution must take this threat into account. The aforementioned scheme has another drawback — it is not resilient against an AFN capture. In Section 7.6, we describe the full-fledged version of SECK that provides effective solutions for both types of attacks (i.e., colluding multiple MSN captures and AFN captures). In the next section we describe the threat model that was considered when designing SECK.

7.5 The Threat Model

We consider an attack scenario where an adversary is able to compromise one or more nodes of a WSN. Specifically, we consider three different cases with differing degrees of severity. Throughout the remainder of the chapter, when stating that a node has been compromised or captured, we assume that all key information held by that node is revealed to the attacker. In the first scenario, an adversary may be able to target and capture an AFN. A less severe attack would be when an attacker captures MSNs within the same cluster. In the third scenario, incurring the least degree of damage, an adversary would simply capture nodes at random throughout the network. One threat that has not been addressed in group key management schemes using administrative keys is the realistic possibility that multiple nodes may be captured before any node capture is detected. Researchers have pointed out that there really is no sure and efficient way to readily detect a node capture [4, 13]. Therefore, for a key management solution to be truly effective in a hostile environment, it must recover from multiple node captures.

The attack scenario that possesses the greatest threat to the bottom tier of the network is the compromise of an AFN, that is, the first attack scenario. This requires an adversary to locate and visually distinguish an AFN from a MSN. Then an adversary must extract the sensitive contents of the AFN (e.g., keys). If an AFN capture is not immediately detected, all data collected by MSNs in that cluster will be compromised. After the detection of the AFN capture, the following steps must be executed to restore normal operations of the cluster: (1) notify MSNs of the capture, (2) establish a new AFN for each MSN, and (3) establish a new security relationship between the MSN and the AFN in the second step. Note that in the second step, MSNs are "reclustered" or absorbed into other existing clusters in their vicinity. If reclustering is not supported by the network, all MSNs within the affected cluster are considered off-line until a new AFN is deployed. The AFN that is captured contains a full set of administrative keys that will need re-keying. To localize the necessary re-keying operations, it is necessary for administrative keys to be independently replaced. If the administrative key sets were globally calculated and distributed to all AFNs, all keys for all clusters would be compromised as a result of a single AFN capture.

In the second attack scenario, MSNs within the same cluster are compromised. In the basic version of SECK described in Section 7.4, it is possible for an adversary to capture not only some, but all of the administrative keys of a cluster with only a few MSN captures. This is possible if the adversary is able to pick the nodes in a strategic manner. In the running example of Section 7.4, the strategic choice would be to compromise N_1 and N_6, which would reveal all five administrative keys. Therefore, if the detection of node captures is not possible, or not prompt, the entire cluster will likely be rendered insecure unless a method of recovery is developed. SECK provides a recovery method to salvage uncompromised MSNs within a cluster when some or even all administrative keys are compromised.

In the third attack scenario, an attacker may compromise nodes randomly throughout the network. A clustered architecture's main advantage, in terms of security, is its ability to localize the effects of attacks on randomly chosen nodes. More discussions on this topic are given in Section 7.7.1. A secondary advantage is that multiple decentralized attacks may not have increased effect compared to a single attack instance. If two adversaries located randomly throughout the network compromise one node each, combining the information obtained from these nodes provides no added benefit, assuming the nodes are not within the same cluster. Of course, this is true only if each AFN generates its administrative keys independently from others.

7.6   The Complete Specification of SECK

We have shown how SECK maintains fresh keys and revokes single users from secure communication. We now describe the mechanisms needed to complete SECK. We start our discussions by describing the full set of keys stored by each MSN to support SECK. We also describe a *location-training* scheme that sets up the network clusters and the coordinate system used to establish administrative keys. Then we describe techniques for replacing administrative keys compromised by multiple MSN captures. Next we describe a reclustering scheme for salvaging MSNs within a cluster after the AFN of that cluster has been captured. Finally, we describe a method to dynamically add an MSN to the network. In Figure 7.2 we show the overall operation and components that constitute SECK.

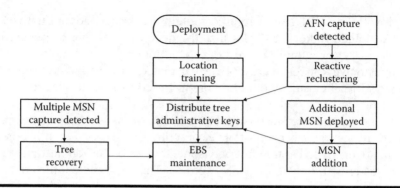

Figure 7.2   Components of SECK.

7.6.1   Required Keys

Initially each MSN will be deployed with the complete administrative key set $\{Ka_1, Ka_2, \ldots, Ka_{k+m}\}$. After a short setup period, only a subset of these keys will be stored. In addition, each MSN will be required to store key Kp_i, which is a pair-wise secret key shared with the BS, and one tree administrative key Kt_i. Kp_i is needed for the reclustering process after the capture of the AFN has been detected. The key Kt_i is used to replace compromised administrative keys after MSN captures have been detected.

7.6.2   Location Training

The location-training scheme will establish a cluster of MSNs around a primary AFN, AFN_p, and assign each MSN a cluster coordinate identifier. This identifier is needed to establish each MSN's administrative key subset, and is also used when routing data to an MSN's primary AFN. In addition, this scheme enables an MSN to store the location of the next-hop MSN, N_x, in the direction of its backup AFN, AFN_b. Each MSN is absorbed into AFN_b's cluster in the event of AFN_p's compromise or failure. A cluster coordinate established is given as (*tree, hopcount*), where *tree* is an integer assigned by AFN_p, and *hopcount* is the MSN's distance from AFN_p. We define a *tree* as a set of MSNs that routes packets through the same tree root when forwarding data to the AFN_p, where the tree root is an MSN that is one hop away from the AFN_p. Once a cluster coordinate has been established for each MSN, that coordinate is used to establish the corresponding key subset from the stored global administrative key set.

It is assumed that at this point MSNs have completed neighborhood discovery, and every MSN and AFN is aware of the unique identities (IDs) of all one-hop neighbors through broadcasted "hello" messages (we assume each MSN and AFN is embedded with a unique ID before deployment). Now, each AFN broadcasts the list of one-hop neighbors that it has discovered to all MSNs within each transmission range. Each entry in the list is a tuple of an MSN ID and its assigned tree number (assignment of tree numbers is described in Section 7.6.3). This broadcast transmission can be expressed as

$$AFN_i \Rightarrow N_1, \ldots, N_m :$$

$$ID_{AFN_i} \parallel (ID_1, tree_1) \parallel (ID_2, tree_2) \parallel \ldots \parallel (ID_{|S_{b1}|}, tree_{|S_{b1}|}),$$

where m denotes the number of MSNs within the transmitting range of the AFN, and S_{b1} denotes the set of MSNs within one hop of AFN_i. Each one-hop node will serve as a tree root for multi-hop nodes established in that tree. MSNs search this list for their ID and the ID of their discovered neighbors. If a node finds its ID on this list, it assigns its cluster coordinate as $(tree_{ID}, 1)$ and becomes one of the tree roots of AFN_p. If an MSN does not find its ID, but finds the ID of a neighbor, say i, it assigns itself the cluster coordinate $(tree_{ID_i}, 2)$ — the first entry corresponds to the tree number of the neighbor MSN and the second entry indicates the hop count from AFN_p. MSNs with multiple neighbors on the neighbor list of AFN_p should randomly choose which tree to join.

The group of second-hop neighbors, S_{b2}, initiates the propagation of the coordinate establishment message by broadcasting their cluster coordinate and ID_{AFN_p} as follows:

For all $N_i \in S_{b2}$, $N_i \Rightarrow$ neighbors : $\qquad ID_{AFN_p} \parallel (tree, hopcount)_{N_i}$

Upon hearing this message, MSNs not yet holding a cluster coordinate will know how many hops away from AFN_p they are, and what their cluster coordinate should be. As these messages propagate, all MSNs will establish their cluster coordinates. An MSN will always forward the first coordinate establishment message it receives. Once an MSN begins to receive additional coordinate establishment messages, it will forward this message only if it is the closest backup AFN heard so far, that is, if $hopcount_{new} < hopcount_b$, where $hopcount_b$ is the number of hops to AFN_b. Each MSN will store ID_{AFN_b}, $hopcount_b$, and the MSN ID of the node that sent the coordinate establishment message containing ID_{AFN_b}. This MSN will serve as N_x toward AFN_b. We describe, in Section 7.6.5, the importance of establishing N_x and AFN_b. Every MSN, with the exception of the $|S_{b1}|$ established root nodes, must broadcast one message when establishing their primary cluster coordinate. Every MSN must transmit at least one additional message when establishing AFN_b. We will assume the case where the optimal AFN_b is established with the first coordinate establishment message received. Then, the total communication overhead for the location training process

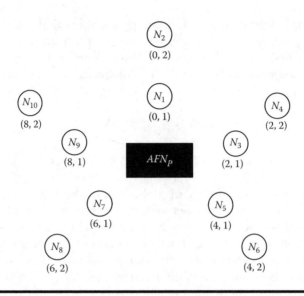

Figure 7.3   Optimal cluster coordinates established in a single cluster.

is $1 + 2n - |S_{h1}|$ transmissions per cluster, where *n* denotes the number of MSNs in the cluster. Using our running example, Figure 7.3 shows cluster coordinates established within a ten MSN cluster. This figure shows the optimal case, because each node receives a unique cluster coordinate. In the following section we explain how the sum of the entries in each node's cluster coordinate yields a unique subset identifier between 1 and 10.

7.6.3   Establishment of Administrative Keys

The administrative key subset is determined by the assignment of the AFN cluster coordinate. We propose an algorithm for assigning tree numbers based on the expected number of hops within a cluster. We assume that the expected number of hops can be estimated based on the density of AFN deployment relative to MSN deployment. We denote the number of neighbors found by an AFN (or degree of AFN) as d, $(d = |S_{h1}|)$ and the maximum number of estimated hops within a cluster as h. We assign tree numbers as $tree_i = i \cdot h$ for $0 \leq i < d$. The administrative key subsets are then determined by the sum of the two elements of a node's cluster coordinate. Because a cluster coordinate consists of *(tree,hopcount)*, an MSN's key subset identifier is calculated as *tree + hopcount*. Each unique cluster coordinate within the expected maximum number of hops within a cluster, h, will generate a unique key subset identifier in the range of $[1, d \cdot h]$.

Each AFN is responsible for generating and distributing the tree administrative keys for each tree in its cluster. This requires $d \cdot h$ messages transmitted by the AFN, one for each of the unique key subset identifiers used in the cluster. The first step is to generate d tree administrative keys, where Kt_j is the tree administrative key for all nodes in the j-th *tree*. Recall that a *tree* consists of all nodes that utilize the same forwarding route. From Figure 7.3, we see that N_1 and N_2 are members of *tree*$_0$. Each Kt_j is distributed to each N_i in *tree*$_j$ as follows:

$$\text{For all } N_i \in tree_j, \text{ AFN} \rightarrow N_i: \qquad \mathbf{E}_{Ka_1}(\mathbf{E}_{Ka_2}(\dots \mathbf{E}_{Ka_{|S_i|}}(Kt_j)\dots))$$

where $\rightarrow$ represents unicast transmission and $S_i = \{Ka_1, Ka_2, \dots, Ka_{|S_i|}\}$. It is important for the administrative key set, U, to be independently updated within each cluster after all tree administrative keys have been established. If the same set of administrative keys is maintained globally, the compromise of a single cluster's keys would compromise the entire network's keys. We described in Section 7.5 that in order to isolate an attack, the administrative keys must be independently generated in each cluster and not shared among clusters.

7.6.4 Administrative Key Recovery

If all administrative keys have been compromised due to multiple node captures within an isolated set of trees, then the remaining trees can be salvaged using their tree administrative keys to reestablish their administrative key subsets.

In Section 7.5 we stated that the capture of a group of MSNs can compromise most or all of the administrative keys in the basic version of SECK. When all of the administrative keys of the cluster have been compromised, even the MSNs that are not captured are excluded from any further communications with the rest of the network. Hence, it is necessary to distribute new session keys to those MSNs. We assume a scenario where an AFN at some point receives notification that a set of nodes, which we denote as S_c, has been compromised, resulting in the compromise of all administrative keys. In response, the AFN computes the set of trees not containing any node from S_c, which we denote as S_{ut}, as

$$S_{ut} = \{tree_i : tree_i \cap S_c = \emptyset, \forall tree_i\}$$

The AFN then creates $|S_{ut}|$ messages, each containing a set of new administrative keys, and transmits the messages to the appropriate trees as shown in the following expression:

$$\text{For all } tree_i \in S_{ut}, \text{ AFN} \Rightarrow tree_i:$$

$$\mathbf{E}_{Kt}(\mathbf{E}_{Ka_1}(Kd_1') \parallel \mathbf{E}_{Ka_2}(Kd_2') \parallel \dots \parallel \mathbf{E}_{Ka_{k+m}}(Kd_{k+m}'))$$

Note that the technique described above cannot salvage MSNs that belong to a tree in which some of its nodes have been compromised. In Section 7.7.3, we show that the technique described above can salvage a greater percentage of MSNs when the attack is more localized (i.e., concentrated in a specific region of a cluster).

7.6.5 Reactive Reclustering after AFN Capture

Unlike an MSN, an AFN carries out several key tasks that are essential to its cluster. Because the loss or capture of an AFN can incapacitate the entire cluster, giving the MSNs the ability to recover from such a situation greatly improves the survivability of the cluster by removing the single point of failure [15]. To keep the MSNs operational after an AFN capture, we need to recluster the MSNs into neighboring clusters and establish new security relationships between the reclustered MSNs and their respective backup AFNs. We identify two approaches to address this problem. First, MSNs may proactively maintain backup security relationships with neighboring AFNs, similar to the approach in Gupta and Younis [16]. This requires the MSNs to participate in periodic key update procedures with multiple AFNs. The second approach involves MSNs reactively utilizing a trusted third party (TTP) to establish a new security relationship with a backup AFN after the capture of the primary AFN is detected. It is reasonable to assume that, in mission-critical applications, the AFNs would be equipped with tamper-resistant hardware because of their importance. We assume that successful AFN captures are infrequent events and adopt the reactive approach because it does not require the MSNs to maintain backup information.

In SECK's approach, the BS acts as the TTP. The BS establishes a security relationship between the MSN (that is being reclustered), N_i, and AFN_b. The BS authenticates AFN_b and N_i, and distributes a new pair-wise key to those nodes so that a new administrative key subset may be established.

Before describing the recluster procedure, we make some prefatory comments. We assume that the AFNs are each equipped with a high-end embedded processor (e.g., Intel Xscale PXA250) that is capable of executing asymmetric cryptographic operations. Recall that at the completion of the location training procedure, each MSN is aware of the identities of the AFN_B and the next-hop MSN en route to the AFN_b, N_x.

Upon detection that N_i's primary AFN has been captured, N_i constructs a short recovery request message destined for the BS. Then N_i sends the following message to N_x.

$$N_i \rightarrow N_x : \qquad ID_{N_i} \parallel ID_{AFN_b} \parallel nonce_i \parallel MAC_{Kp_i}$$

where MAC_{Kp_i} represents a message authentication code (MAC) generated with N_i's base station pair-wise key, Kp_i. We assume that N_x routes this message to AFN_b in the same manner as forwarding sensed data to AFN_b for data aggregation. After receiving this message, AFN_b digitally signs this message, appends the signature to the message, and forwards the following message to the BS:

$$AFN_b \rightarrow \text{BS}: \qquad ID_{N_i} \parallel ID_{AFN_b} \parallel nonce_i \parallel MAC_{Kp_i} \parallel Sign_{AFN_b}$$

where $Sign_{AFN_b}$ represents AFN_b's signature. When the BS receives this message, it authenticates N_i using the MAC and then authenticates AFN_b using the signature. If both nodes are authenticated, the BS generates a secret key $K_{AFN_b-N_i}$ and transmits it to both AFN_b and N_i as shown in the following equations:

$$\text{BS} \rightarrow AFN_b:$$

$$ID_{N_i} \parallel ID_{AFN_b} \parallel \mathbf{E}_{K_{AFNb+}}(K_{AFN_b-N_i} \parallel ID_{N_i} \parallel ID_{AFN_b} \parallel nonce)$$

$$\text{BS} \rightarrow N_i:$$

$$ID_{N_i} \parallel ID_{AFN_b} \parallel \mathbf{E}_{Kp_i}(K_{AFN_b-N_i} \parallel ID_{N_i} \parallel ID_{AFN_b} \parallel nonce)$$

where K_{AFNb+} represents the public key of AFN_b. Using $K_{AFN_b-N_i}$, N_i and AFN_b can now establish a security relationship. Using $K_{AFN_b-N_i}$, AFN_b finalizes N_i's membership in its cluster by sending N_i a subset of administrative keys and the corresponding subset identifier. It is noted that if an EBS reaches its maximum number of nodes, that is, $n = {}_{k+m}C_k$, adding a new node will require the expansion of the EBS by adding a new key. For brevity, we do not describe the process of extending an EBS in this chapter; we simply assume that the size of a cluster's initial EBS will be sufficient for node additions and refer the reader to Eltoweissy et al. [10] for an EBS expansion mechanism.

7.6.6 MSN Addition

Throughout the lifetime of a WSN, it may be necessary to deploy additional MSNs. We propose a way of adding nodes to an existing WSN. We assume that MSNs are randomly deployed and that the MSN's resulting cluster is unknown. In SECK, each cluster maintains a unique and private set of keys. For this reason it is impossible for a new MSN to be predeployed with any keys that will enable authentication with a MSN or AFN directly. However, each new MSN will contain a unique base station pair-wise key, Kp_i, which allows the BS to facilitate a new security relationship between the new MSN and existing AFN in the same way as the reclustering process described in Section 7.6.5. The only difference between deploying a new MSN and reclustering an MSN is that the newly deployed MSN has no knowledge of a backup AFN. Once a newly deployed MSN knows which AFN's cluster to join, it starts the reclustering process described above.

To determine which AFN's cluster is most appropriate to join, a new MSN, N_{new}, conducts a survey of neighbor nodes' cluster status. First, N_{new} broadcasts a short "hello" message to announce its presence. Each neighbor, $neighbor_i$, that overhears this message replies with its cluster information as follows:

$$neighbor_i \rightarrow N_{new}: \qquad ID_{N_i} \parallel ID_{AFN_p} \parallel hopcount_{N_i}$$

where ID_{AFN_p} is $neighbor_i$'s primary AFN, and $hopcount_{N_i}$ is the hopcount in $neighbor_i$'s cluster coordinate identifier. N_{new} uses these replies to determine the most appropriate cluster to join and the most efficient neighbor to use as a next-hop node to that cluster's AFN. N_{new} determines the most appropriate AFN, AFN_{max}, based on the number of neighbors belonging to AFN_{max}. N_{new} selects the most efficient neighbor based on the neighbor's $hopcount$ to AFN_{max}. The neighbor with the smallest $hopcount$ is the most efficient node to use as a next-hop node in the direction of AFN_{max}. Now that the next-hop node to AFN_{max} is known, N_{new} conducts the reclustering procedure described in Section 7.6.5.

7.7   Robustness of SECK against Node Capture

Node capture in hostile environments is inevitable, and an effective key management scheme should be able to recover from such attacks to be effective. We describe some of the inherent security advantages of utilizing a clustered and hierarchical network architecture. Then, using the threats identified in Section 7.5, we analyze how well SECK recovers from those attacks.

7.7.1   Robustness of a Clustered Architecture

A clustered and hierarchical framework for WSNs provides many beneficial security properties. Isolation is the primary benefit of a clustered key management scheme. Each AFN is responsible for independently calculating and periodically distributing new administrative keys. Hence, an attack that yields keys within one cluster will not impact any other cluster in the network. An adversary must perform a global attack to completely compromise the network. This is not the case for most nonhierarchical key management schemes. In nonhierarchical key management solutions, a localized attack often has a global impact on the network's keys [3,17,19,21].

7.7.2 Robustness against MSN Node Capture

In the example given in Table 7.3, it is possible for an adversary to capture all the administrative keys of a cluster with only a small number of node captures by strategically selecting the nodes to capture. For example, the capture of N_1 and N_6 would reveal all five administrative keys. Fortunately, such attacks are difficult to carry out; to be successful, an adversary needs to know which administrative keys are stored in each MSN. Of course, the adversary can always randomly choose the nodes to capture and hope that a large proportion of administrative keys is revealed by those captures. Using simulations, we show that when the nodes are compromised randomly, then the proportion of compromised keys is commensurate with the well-known probabilistic key distribution scheme proposed in Eltoweissy and Gligor [3].

Eschenauer and Gligor (EG) [3] proposed a probabilistic key establishment technique for WSNs in which pair-wise secret keys are selected from a global key pool. In the past few years, similar schemes have been proposed [19, 21]. The EG scheme has been shown to have acceptable key resiliency properties. That is, the capture of a limited number of nodes does not reveal much key material of other nodes. Figure 7.4 compares three instances of SECK with the EG scheme by plotting their ratio of keys captured as a function of ratio of nodes captured. For SECK, the keys being considered are administrative keys; for EG, the keys being considered are pair-wise

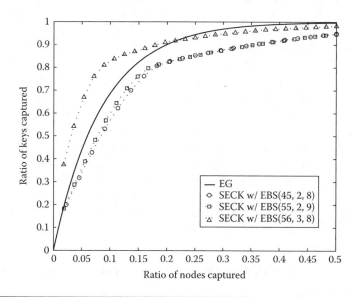

Figure 7.4 Expected ratio of keys captured versus ratio of nodes captured.

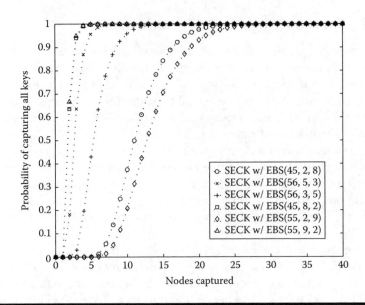

Figure 7.5 Probability that all keys have been compromised versus the number of nodes captured.

secret keys. Figure 7.4 shows that SECK's robustness against random node (MSN) captures is commensurate with that of the EG scheme.

Now we attempt to answer the following question: how many MSNs must be captured before all the administrative keys of a cluster ($k + m$ keys in total) are compromised? In Figure 7.5, the probability of capturing all the administrative keys as a function of the number of nodes captured is plotted for several instances of SECK, with all instances supporting approximately 50 nodes. We are interested in instances of SECK that support about 50 nodes, as that is the expected size of a typical cluster that we envision. The figure shows that the ratio k/m has a direct impact on the robustness of SECK against node captures — decreasing the ratio improves robustness against node captures and increasing the ratio has the opposite effect. However, setting the ratio k/m to a low value incurs a cost — as the ratio decreases, the communication overhead required to distribute session keys increases. One can see that there exists a communication overhead versus node capture resiliency trade-off. Further discussions on this issue are continued in the next section.

7.7.3 Evaluation of the Administrative Key Recovery Procedure

Recall that SECK generates d tree administrative keys for a cluster consisting of n MSNs in which every MSN is within b hops of its primary AFN.

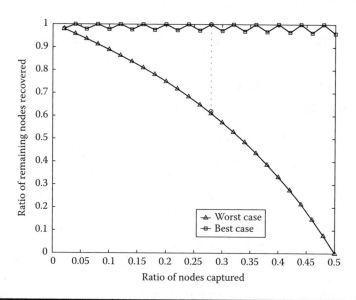

Figure 7.6   Administrative key recovery procedure evaluation.

Figure 7.5 shows that SECK, using EBS(55,2,9), provides the most resiliency against node captures. If 14 MSNs have been randomly captured in this case, there is a good chance (>60 percent) that the complete administrative key set has been compromised. If all the administrative keys of a cluster are compromised, the network needs to execute the MSN administrative key recovery procedure. In the following paragraphs we discuss the effectiveness of the administrative key recovery procedure in two distinct cases — best and worst case scenarios — assuming that x MSNs have been captured.

The worst case occurs when each of the x captures occurs in separate trees. This leaves $d - x$ trees unaffected, and $(d - x) \cdot h$ nodes can be recovered. If we approximate d with n/h, the ratio of nodes that can be recovered is $(n - x \cdot h)/(n - x)$. Suppose that every MSN in the cluster is within two hops of the AFN (i.e., $h = 2$) and EBS(55, 2, 9) is used. Then our procedure recovers 0.66 of the uncompromised nodes in the worst case.

The best case occurs when the attack is completely localized. That is, all nodes within a single tree are captured before the attacker moves on to the next tree. This will affect $\lceil x/h \rceil$ trees, leaving $d - \lceil x/h \rceil$ trees unaffected. If we again approximate d with n/h, the ratio of nodes that can be recovered is $(n - h \cdot \lceil x/h \rceil)/(n - x) \approx 1$. From the above analysis we can observe that SECK's recovery procedure performs best in localized attacks.

In Figure 7.6, we have plotted the ratio of recoverable nodes in the best and worst case attack scenarios. We assume that EBS(55, 2, 9) is employed. In the figure, we have highlighted the case where 14 nodes have been captured. In practice, the actual recovery ratio is expected to be somewhere between 0.6 and 1.0 when 14 nodes are captured. As expected, as the ratio of nodes captured increases, the ratio of recoverable nodes decreases.

7.8 Evaluation of Communication and Storage Overhead

7.8.1 Energy Dissipation

In typical key management schemes, the energy required for computation is three orders of magnitude less than that required for communication [15]. Moreover, the amount of energy consumed for computation varies significantly with hardware. Hence, we only consider energy costs associated with radio signal transmission and reception, and do not consider energy costs associated with computation. To calculate the amount of energy dissipated during the execution of SECK, we use the power usage specification of Sensoria's WINS NG: the RF radio component consumes 0.021 mJ/bit for transmission, and 0.014 mJ/bit for reception when operating at 10 kbps [26]. We make the same assumptions as in Carman et al. [15] with regard to the following message element sizes:

- All node IDs are 64 bits.
- All nonces are 64 bits.
- All symmetric keys are 128 bits.
- The *tree* identifier is 32 bits.
- The *hopcount* is 32 bits.
- All MACs are 128 bits.
- The RSA modulus used for digital signatures is 1024 bits.

Table 7.4 shows the amount of energy dissipated by a single node to complete one instance of each process (six processes are listed). We assume that EBS(55,2,9) is employed in a cluster consisting of 50 MSNs with $b = 2$. Each AFN has a neighbor degree of 20; that is, $|S_{b_1}| = 20$. In the following we discuss the major factors that contribute to the energy consumption for each of these processes.

The energy required for the location training process is affected by the efficiency at which the MSNs find their optimal backup AFN. We assume that the backup AFN is found when a node receives the second coordinate establishment message. To distribute the tree administrative keys to all MSNs in its cluster, an AFN must generate one message for each MSN in its cluster. Hence, the energy consumed by each AFN is directly proportional to the number of MSNs in its cluster. The communication overhead required for the EBS maintenance process is the message transmission needed for refreshing all administrative keys, and therefore this overhead depends on the dimension of the EBS that is used. The energy cost of the tree recovery

Table 7.4   SECK Communication Energy Consumption (in mJ)

		Transmission	Reception	Total
Location training	AFN	55.10	N/A	55.10
	MSN	1.10	36.74	37.84
Distribute tree	AFN	336.00	N/A	336.00
administrative key	MSN	N/A	4.48	4.48
EBS maintenance	AFN	44.35	N/A	44.35
	MSN	N/A	29.56	29.56
Tree recovery	AFN	29.57	N/A	29.57
	MSN	N/A	19.71	19.71
Reactive	AFN	38.98	11.65	50.63
reclustering	MSN	17.47	11.65	29.12
MSN addition	AFN	38.98	11.65	50.63
	MSN	20.67	11.65	32.32

process shown in Table 7.4 is the energy dissipated by the AFN for each *tree* recovered. The energy cost of reactive reclustering calculated in Table 7.4 is the energy needed to forward the recovery messages needed to recover one MSN. Therefore, the total energy dissipated during the cluster recovery process depends on the number of MSNs being recovered with a specific backup AFN. Finally, the energy cost during node addition is calculated as the energy needed to respond to, then to forward the recovery messages of one newly added MSN. It can be seen from Table 7.4 that SECK effectively offloads much of the energy-intensive operations to the more capable AFN.

7.8.2   Storage Overhead

Prior to deployment, each MSN needs to store a key subset matrix and the complete administrative key set. The key subset matrix is a $(k+m) \times n$ bit matrix that identifies the keys associated with each subset. Table 7.3 is a sample 5×10 key subset matrix. The key subset matrix requires $n \cdot (k+m)$ bits to store, and the complete administrative key set requires $128 \cdot (k+m)$ bits to store (here, we assume that 128-bit AES keys are used). Additionally, one 128-bit base station pair-wise key is stored at each MSN, for a total initial storage requirement of $((128 + n) \cdot (k+m) + 128)$ bits. With this formula, one can calculate that each MSN would need to store 266 bytes of initial keying material if EBS(55,2,9) is employed.

After deployment, each MSN deletes most of its initial keying material immediately after the conclusion of the location training processes. Specifically, each MSN deletes its key subset matrix and the unused administrative keys. Recall that each MSN is assigned one 128-bit tree administrative key after deployment. Assuming that EBS(55,2,9) is employed, each MSN would need to store 64 bytes of keying material or four 128-bit keys at the end of the location training process.

7.8.3 Comparison of Communication Overhead

The keying communication overhead of SECK is incurred during (1) the initial key establishment phase and (2) periodic key maintenance procedures. We compare the communication overhead incurred in (1) and (2) with those incurred in the Localized Encryption and Authentication Protocol (LEAP) [13] and Simple Key Distribution Center (SKDC) [15], respectively. A unique feature of SECK is its use of both administrative keys and session keys. Because of this feature, we cannot compare SECK, in its entirety, with a single scheme. Instead, we decompose SECK into (1) and (2) and compare the two constituent parts with LEAP and SKDC that respectively carry out similar functions. LEAP supports multiple levels of communication through multiple degrees of key sharing. SECK provides a similar level of flexibility within a clustered architecture. SKDC is a session key distribution scheme that utilizes a leader node to distribute the session key. It was shown in Carman et al. [18] that this basic method used by SKDC is the most efficient means to distribute a session key.

7.8.3.1 Key Establishment

In this subsection we compare the communication overhead incurred by SECK during the key establishment process with that of LEAP. LEAP is a well-known key management scheme that provides communication flexibility by establishing multiple classes of keys. In SECK, a location-training process is executed to establish administrative key sets; then a distribution process is executed to distribute a session key. LEAP goes through similar key establishment processes to establish each node's pair-wise and cluster keys.

LEAP restricts each node to three secure communication groups. SECK places no such restriction and provides a simple mechanism to establish secure communication groups of any degree within a cluster. In LEAP, every MSN has a distinct pair-wise key established with each neighbor. In SECK, however, every MSN establishes a pair-wise key only with its primary AFN.

In LEAP, each node calculates a single pair-wise key to share with each neighbor. This calculated pair-wise key is unicasted to each of the node's d neighbors. In addition, establishing a LEAP cluster key requires $n(d-1)^2/(n-1) \approx (d-1)^2$ key transmissions throughout the network [13]. LEAP does not provide a concrete message format. To compute the energy dissipation incurred by LEAP, we assume that LEAP has the same message format as that used in SECK for key distribution. Here, we only consider communication-related energy dissipation. In Figure 7.7 we compare SECK

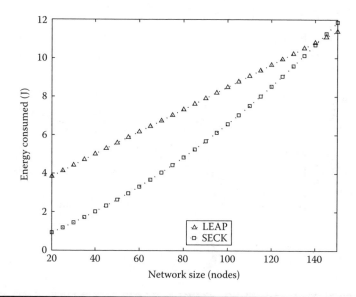

Figure 7.7   Key establishment communication overhead.

and LEAP by plotting the energy dissipated by the network for key establishment as a function of the network size (for SECK, network size implies cluster size). Note that the size of a network (i.e., number of nodes) has the biggest impact on the energy required to establish keys. We set the connection degree as 20, which was suggested by the authors of LEAP.

It is shown in Figure 7.7 that SECK is more efficient for small network sizes. Distributing similar amounts of keying material using the method described in SECK is better than that of LEAP when fewer nodes are considered. For the clustered network architecture described in Section 7.3, we assume that clusters consist of approximately 50 nodes — for such network sizes, SECK outperforms LEAP.

7.8.3.2   Updating Session Keys

We now switch our attention to the communication overhead incurred by SECK to maintain session keys. Carman et al. [18] show that the straightforward technique of unicasting a session key to each group member incurs the least amount of communication overhead among session key distribution schemes. SKDC is a simple instance of such a method and is most effective for a single instance of session key distribution. Our emphasis here is on the efficiency of session key distribution over multiple key update periods.

For a network deployed in a hostile environment, where node captures are expected, it is important to be able to continually distribute new session

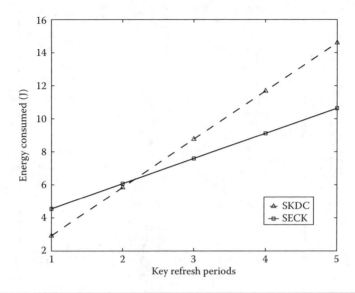

Figure 7.8   Key refresh communication overhead.

keys efficiently. SKDC creates an individual message for each group member every time a session key must be updated, while SECK requires one message for each administrative key. In Figure 7.8 we compare SECK and SKDC in terms of the communication energy dissipated by the network due to session key redistributions. The plot shows that SKDC outperforms SECK during the initial session key update periods. This is because of SECK's initial overhead for establishing the administrative keys. However, SECK outperforms SKDC as the accumulated number of session key redistributions increases. This result was expected — the administrative keys employed in SECK reduce the cumulative number of messages that must be transmitted by the AFN for session key redistributions.

7.9   Related Work

In the past few years, several technical approaches have been proposed to provide WSNs with confidentiality and authentication services via pair-wise secret (or key) sharing [3,13,19,21]. As stated earlier, Eschenauer and Gligor [3] propose a static technique for probabilistic distribution of pair-wise keys. In their scheme, keys are randomly chosen from a global key pool. Chan et al. [19] extended this idea to provide localized attack resiliency. Most WSN applications require additional levels of key sharing beyond pair-wise shared keys — specifically, group keys.

Carman et al. [18] conducted a comprehensive analysis of various group key schemes. The authors concluded that group size is the primary factor that should be considered when choosing a scheme for generating and distributing group keys in an WSN. Zhu et al. [13] proposed a comprehensive key management scheme called LEAP that establishes multiple keys for supporting neighborhood as well as global information sharing. Although LEAP includes several promising ideas, it does not adequately address scalability issues concerning the distribution and maintenance of group keys.

To address the difficult problem of scalability, many have proposed hierarchical network architectures, similar to the one described in this chapter. In [4,5,11], the authors utilize a clustered and hierarchical network architecture for key management. Jolly et al. [4] employ a hierarchical network organization to establish *gateway-to-sensor* keys. The clustering technique used by Jolly et al. was originally developed by Gupta et al. [22]. By using GPS signals, Gupta et al. propose to form a cluster in which all nodes are within one hop of the cluster head. Eltoweissy et al. [5,11] present another hierarchical key management scheme based on the Exclusion Basis System to efficiently maintain group and session key information. This approach supports key recovery only if node captures can be immediately detected.

Bohge et al. [2] propose a hierarchical authentication technique to establish and recover keys. Their approach requires a broadcast authentication scheme, and they employ a variation of μTESLA [24] for this purpose.

7.10 Conclusion

In a large-scale WSN deployed in a hostile environment, a key management scheme is needed to manage the large number of keys in the system. In this chapter we described a cluster-based dynamic key management scheme, SECK, that is designed to address this specific issue. SECK meets the stringent efficiency and security requirements of WSNs using a set of administrative keys to manage other types of keys such as session keys. SECK is a key management solution that includes (1) a location training scheme that establishes clusters and the cluster coordinate system used in the MSN recovery procedure; (2) a scheme for establishing and updating administrative keys; (3) a scheme for distributing session keys using administrative keys; (4) a scheme for recovering from multiple node captures; and (5) a scheme for reclustering and salvaging MSNs in the event that their AFN has been captured. Through analytical and simulation results, we have shown that SECK is resilient to node and key captures while incurring low levels of communication overhead.

References

1. M. Chorzempa, J. Park, and M. Eltoweissy, "SECK: Survivable and efficient clustered keying for wireless sensor networks," in *Proc. IEEE Workshop on Information Assurance in Wireless Sensor Networks,* pp. 453–458. Phoenix AZ, April 2005.

2. M. Bohge and W. Trappe, "An authentication framework for hierarchical ad hoc MSN networks," in *Proc. of the ACM Workshop on Wireless Security (WiSe),* pp. 79–87. San Diego CA, September 2003.

3. L. Eschenauer and V.D. Gligor, "A key-management scheme for distributed MSN networks," in *Proc. of the 9th ACM Conf. on Computer and Communications Security (CCS),* pp. 41–47. Washington D.C., November 2002.

4. G. Jolly, M. Kusu, and P. Kokate, "A hierarchical key management method for low-energy wireless MSN networks," in *Proc. of the 8th IEEE Symposium on Computers and Communication (ISCC),* pp. 335–340. Turkey, July 2003.

5. M. Eltoweissy, A. Wadaa, S. Olariu, and L. Wilson, "Scalable cryptographic key management in wireless sensor networks," *Journal of Ad Hoc Networks: Special issue on Data Communications and Topology Control in Ad Hoc Networks,* Vol. 3, No. 5, September 2005.

6. W. Heinzelman, "Application-Specific Protocol Architectures for Wireless Networks," Ph.D. dissertation. Massachusetts Institute of Technology, June 2000.

7. C. Intanagonwiwat, R. Govindan, and D. Estrin, "Directed diffusion: a scalable and robust communication paradigm for MSN networks," in *Proc. 6th Conference on Mobile Computing and Networking (MobiCom),* pp. 56–67. Boston, MA, August 2000.

8. S. Madden, R. Szewczyk, M. Franklin, and D. Culler, "Supporting aggregate queries over ad-hoc wireless MSN networks," in *Proc. 4th IEEE Workshop on Mobile Computing Systems and Applications,* pp. 49–58. Callicoon, NY, June 2002.

9. A.Wadaa, S. Olariu, L. Wilson, M. Eltoweissy, and K. Jones, "Training a sensor network," *Mobile Networks and Applications,* Vol. 10, No. 1, pp. 151–168, February 2005.

10. M. Eltoweissy, H. Heydari, L. Morales, and H. Sudborough, "Combinatorial optimizations of group key management," *Journal of Networks and Systems Management,* Vol. 12, No. 1, pp. 30–50, March 2004.

11. M. Eltoweissy, M. Younis, and K. Ghumman, "Lightweight key management for wireless MSN networks," in *Proc. IEEE International Conference on Performance, Computing, and Communications,* pp. 813–818. Phoenix, AZ, April 2004.

12. M. Moharram and M. Eltoweissy, "Key management schemes in sensor networks: dynamic versus static keying," *ACM Workshop on Performance Evaluation of Wireless Ad Hoc, Sensor, and Ubiquitous Networks (PE-WASUN 2005).* Montreal, Canada, October 2005.

13. S. Zhu, S. Setia, and S. Jajodia, "LEAP: efficient security mechanisms for large-scale distributed MSN networks," in *Proc. of the 10th ACM Conference on Computer and Communication Security (CCS),* pp. 62–72. Washington D.C., October 2003.

14. R. Moharrum, M. Mukkamala, and M. Eltoweissy, "CKDS: an efficient combinatorial key distribution scheme for wireless," in *Proc. International Conference on Performance, Computing, and Communications (IPCCC)*, pp. 63–636. Phoenix, AZ, April 2004.

15. D. Carman, P. Kruus, and B. Matt, "Constraints and approaches for distributed MSN network security," Network Associates Inc. (NAI) Labs Technical Report No.00010. September 2000.

16. G. Gupta and M. Younis, "Fault-tolerant clustering of wireless MSN networks," *IEEE Wireless Communications and Networking*, Vol. 3, No. 1 pp. 1579–1584, March 2003

17. B. Matt, "A preliminary study of identity-based, group key establishment protocols for resource constrained battlefield networks," Network Associates Labs Technical Report 02-034. September 2002.

18. D. Carman, B. Matt, and G. Cirincione, "Energy-efficient and low-latency key management for MSN networks," in *Proc. of 23rd Army Science Conference*. Orlando, FL, December 2002.

19. H. Chan, A. Perrig, and D. Song, "Random key predistribution schemes for MSN networks," in *Proc. IEEE Symposium on Security and Privacy*, pp. 197–213. Oakland, CA, May 2003.

20. W. Du, J. Deng, Y.S. Han, S. Chen, and P.K. Varshney, "A key management scheme for wireless sensor networks using deployment knowledge," in *Proc. IEEE INFOCOM'04*. March 2004.

21. D. Liu and P. Ning, "Establishing pairwise keys in distributed sensor networks," in *Proc. ACM Conference on Computer and Communications Security (CCS)*, pp. 52–61. Washington D.C., October 2003.

22. G. Gupta and M. Younis, "Performance evaluation of load-balanced clustering of wireless MSN networks," in *Proc. 10th International Conference on Telecommunications*, pp. 1577–1581. Papeete, French Polynesia, February 2003.

23. D. Estrin, R. Govindan, J. Heidemann, and S. Kumar, "Next century challenges: scalable coordination in sensor networks," in *Proc. 5th International Conference on Mobile Computing and Networks (MobiCom)*, pp. 263–270. Seattle, WA, August 1999.

24. A. Perrig, R. Szewczyk, V. Wen, D. Culler, and J. Tygar, "SPINS: Security protocols for sensor networks," *Wireless Networks*, Vol. 8, No. 5, pp. 521–534, November 2002.

25. G.J. Pottie and W.J. Kaiser, "Wireless integrated network sensors," *Communications of the ACM*, Vol. 43, No. 5, pp. 51–58, May 2000.

26. Sensoria Corporation, "WINS NG Power Usage Specification: WINS NG 1.0." January 2000. Available: http://www.sensoria.com/

27. J. Chou, D. Petrovis, and K. Ramchandran, "A distributed and adaptive signal processing approach to reducing energy consumption in sensor networks," in *Proc. IEEE INFOCOM*, Vol. 2, pp. 1054–1062. San Francisco, CA, April 2003.

28. W. Heinzelman, A. Chandrakasan, and H. Balakrishnan, "An application-specific protocol architecture for wireless microsensor networks," in *Proc. 33rd Hawaii Int. Conference on System Sciences*, pp. 3005–3014. Maui, HI, January 2000.

29. C. Karl and D. Wagner, "Secure routing in wireless sensor networks: attacks and countermeasures," *Ad Hoc Networks Journal,* Vol. 1, No. 1, pp. 293–315, January 2003.

30. A.D. Wood and J.A. Stankovic, "Denial of service in sensor networks," *IEEE Computer,* Vol. 35, No. 10, pp. 54–62, June 2002.

31. C.K. Wong, M. Gouda, and S. Lam, "Secure group communications using key graphs," in *Proc. IEEE Transactions on Networking,* Vol. 8, No. 1, pp. 16–29, February 2000.

32. H. Harney and C. Muckhenhirn, "Group Key Management Protocol (GKMP) Specification," RFC 2093, July 1997.

33. Y.T. Hou, Y. Shi, and H.D. Sherali, "Rate allocation in wireless sensor networks with network lifetime requirement," in *Proc. ACM International Symposium on Mobile Ad Hoc Networking and Computing (MobiHoc),* pp. 67–77. Tokyo, Japan, May 2004.

34. J. Pan, Y.T. Hou, L. Cai, Y. Shi, and S.X. Shen, "Topology control for wireless sensor networks," in *Proc. ACM International Conference on Mobile Computing and Networking (Mobicom),* pp. 286–299. San Diego, CA, September 2003.

SECURE ROUTING IV

Chapter 8

Scalable Security in Wireless Sensor and Actuator Networks (WSANs)

Fei Hu, Waqaas Siddiqui, and Krishna Sankar

Contents

8.1 Abstract

This chapter discusses the challenging security issues in wireless sensor and actuator networks (WSANs), a special type of wireless sensor networks (WSN). Because WSANs have specific network constraints and data transmission requirements compared to general ad hoc networks and other wireless/wired networks, we propose to seamlessly integrate WASN security with a ripple-zone (RZ)-based routing architecture that is scalable and energy efficient. In this research, we also develop a two-level re-keying/re-routing scheme that is able to not only adapt to a dynamic network topology, but also securely update keys for each data transmission session. Moreover, to provide the security for the in-networking processing such as data aggregation in WSANs, we define a multiple-key management scheme in conjunction with our proposed ripple-zone routing architecture. Extensive simulations and hardware experiments have been conducted to verify the energy efficiency and security performance of our security scheme for WSANs.

8.2 Background

More recently, an important type of network, which is based on the integration of mobile ad hoc networks (MANETs) that consists of mobile *actuators* and wireless sensor networks (WSNs) with large amounts of low-energy,

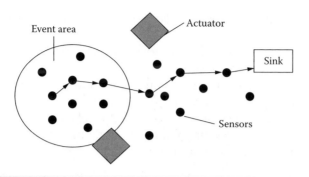

Figure 8.1 WSAN architecture.

tiny *sensors,* have played more and more important roles in homeland security applications [1,2]. Such hybrid networks are usually called *wireless sensor and actuator networks (WSANs)* [2], which cannot be simply regarded as MANETs due to the coexistence of mobile actuators (forming MANET) and fixed sensors (forming WSN). As shown in Figure 8.1, actuators execute corresponding tasks based on the collected sensing data from sensors. Please notice that there are some important differences between the two components in a WSAN (i.e., the MANET and the WSN): (1) the number of nodes in a WSN is significantly larger than in a MANET; (2) sensors are usually low-cost devices with severe constraints with respect to energy source, computation capabilities, and memory, while actuators generally have relatively higher energy storage, which allows longer wireless transmission distance; (3) sensors are usually stationary or with quite limited mobility; and (4) the mode of communication in WSNs typically is many-to-one (from sensors to sink), while it is typically peer-to-peer in MANETs. While WSNs are concerned only with sensor-to-sink interconnections, in WSANs four types of coordination must be considered in the same scenario: actuator-to-actuator (A-A) (to determine which actuators should respond to which sensing area), sensor-to-sensor (S-S) (to use multi-hop communication mode to transmit sensing data), actuator-to-sensor (A-S) (downlink transmission to instruct sensors to execute a certain sensing task), and sensor-to-actuator (S-A) (uplink transmission to report new events or required query results). A-A coordination can be regarded as a MANET issue that has been studied extensively. S-S coordination is a topic of WSN that is still a largely unexplored field [3].

Security is important in many WSAN-based civilian/military applications, such as disaster recovery (earthquake and fire rescue, etc.), airport terrorist-attack prevention, industrial manufacturing control, etc. The study of security should consider the following challenges in WSANs:

1. Consider actuator/sensor coordination and actuator/sensor hetero-geneity.
2. Protocol simplicity (because sensors have limited memory and computational capability [3]).
3. Algorithm scalability (there could be hundreds, if not thousands, of sensors in a typical WSAN application [1]).
4. Low energy consumption (to extend the lifetime of tiny sensors). The energy for one bit of wireless communication can be used to execute over 1000 local instructions in a sensor [2].

Thus, the WSAN trustworthiness protocols should have low wireless communication overhead.

Many of the current sensor network security schemes are based on a key predistribution strategy [6–8]. It works as follows. Before sensor deployment, a subset of key pool is assigned to each sensor to make two sensors likely share a pair-wise key. However, we argue that only key predistribution cannot achieve satisfactory security performance because the attackers can capture some sensors/actuators and learn those permanent keys. It is important to update those keys (i.e., using re-keying) from time to time for different packet transmission sessions. Re-keying is also important in terms of adaptation to network topology changes due to node failure, node addition, and interruptions in the wireless transmission medium [5]. For example, if nodes fail due to low power, messages will fail to be delivered because routes containing the dead nodes still exist. In the case of node addition, it is important to distinguish between legitimate sensor traffic and the infiltration of the network by an enemy node. Also, intermittent connectivity must be considered because the security scheme should be able to deal with wireless errors. These considerations, along with the resource (battery, memory, etc.) constraints imposed, make the design of a WSAN security scheme an extremely difficult task.

The contributions and innovations of this proposed WSAN security scheme include the following four aspects:

1. *Seamless integration of security with scalable WSAN routing protocols*: Our scheme is highly practical because it was designed to integrate routing layer and security protocol without sacrificing power. It is a dynamic, distributed protocol where security is provided independent of central control. Existing work overlooks the idea that any security scheme should be seamlessly integrated with the special characteristics of sensor network architecture, especially routing protocols; otherwise, the security scheme may not be practical or energy efficient from the network protocol point of view [3]. Our security strategy considers special WSAN topology through two-level keying.

Thorough research of the field has found that most of the existing sensor network security strategies focus only on key management/ security algorithms and ignore the specific *routing* characteristics of sensor networks. For example, all existing key-predistribution schemes try to establish pair-wise keys between each pair of nodes. However, most sensors do not need to establish a direct point-to-point secure channel with sensors multiple hops away because sensor network use hop-to-hop communication techniques to achieve long distance transmission. One of the most famous schemes, SPINS [9], simply assumes a flooding-based, spanning-tree architecture with the BS (base station) as the tree-root. However, the establishment and maintenance of a global spanning tree in a large-scale WSN with a large footprint may not only bring unacceptable communication overhead (and thus increased energy consumption[1]), but also cause a large transmission delay and make the assumption of time synchronization in μTESLA (a broadcast authentication protocol [9]) impractical. Another important feature of our work is that it has a robust hop-to-hop transmission scheme and can recover from multiple key losses.

2. *Dynamic security.* Dynamic network topology is native to WSANs because nodes can fail or be added. In the case where nodes fall out, these nodes must be recognized as *dysfunctional* from the viewpoint of the rest of the network. In the case of node addition, a protocol must be able to distinguish between legitimate node addition and attempted enemy infiltration. Given these reasons, *adaptive security* should be present for WSAN applications to ensure overall network security.

3. *Robust re-keying.* From time to time, network enemies might compromise sensors and all security information in those sensors may be obtained. Therefore, after key predistribution and sensor deployment, a *re-keying* scheme should be used to update some types of keys, such as group keys (for broadcast security) and session keys (for securing the current data packets). This is done to ensure that enemies cannot acquire the keys easily. In this work, a re-keying protocol that can adapt to dynamic network conditions such as sensor compromise and topology change is planned.

4. *Low-complexity implementation.* Our work uses a symmetric-key-based scheme because memory use is a major concern in sensors [3]. This prohibits the use of memory-intensive asymmetric keying

---

[1] Most of the sensor's power is consumed by communication, not by instruction processing [3]. The energy used to communicate one bit of data can be used to execute over 1000 local instructions [3]. High energy consumption can drain sensors quickly and thus shorten the network lifetime.

schemes. Asymmetric keying schemes need more complex cryptography calculations and protocols, which can bring too much communication overhead compared to symmetric-key-based schemes. Our security protocol also has low transmission energy due to its cluster-based key management. Because a WSAN typically consists of hundreds, if not thousands, of nodes, network topology/densities can change; therefore, a centralized or flooding-based security scheme cannot scale well. Thus, distributed algorithms and localized coordination to achieve global convergence and scalability are preferred [3].

The remainder of this chapter is organized as follows. In Section 8.2, we point out the shortcomings of the related works and the importance of this work; Section 8.3 first provides our scalable routing architecture and then discusses the security issues in high-level nodes of WSAN (i.e., among actuators). In Section 8.4, security among low-level nodes (among sensors in each actuator domain) is described in detail; extensive simulation results are given in Section 8.5. Section 8.6 is the security analysis; hardware experiments are stated in Section 8.7. Finally, Section 8.8 concludes this chapter.

8.3 Related Works

As pointed out by Akyldiz and Kasimoglu [2], currently, very little research work has been conducted on WSANs that have the coexistence of mobile actuators and low-energy sensors. A contention-free MAC protocol for WSAN is presented in Carley et al. [24]. In Chow et al. [25], WSANs are only examined from the control engineering perspective. While the A-A coordination is investigated in Gerkey and Mataric [26], existing and emerging technologies in WSANs are briefly described without consideration of the interaction between the sensors and actuators in Haenggi [27] and Hu et al. [29].

The closest related work to WSANs occurs in the general wireless sensor networks (WSNs), which is a hot research field today. A survey of the early work on WSNs is provided in Alkyildiz et al. [3]. In terms of security issues in general sensor networks (not WSANs), the pioneering work on securing WSN end-to-end transmission is SPINS [9], which requires time synchronization among sensors. It also proposes μTESLA, an important innovation for achieving broadcast authentication of any messages sent from the base station (BS). An improved multi-level μTESLA key-chain mechanism was proposed in Liu and Ning [30,31]. A key-pool scheme was suggested in Eschenauer and Gligor [10] to guarantee that any two sensors share at least one pair-wise key with a certain probability. Multiple pair-wise keys may be found between nodes by the schemes proposed in Chan et al. [11]. Key predistribution schemes utilizing location information were described in

Liu and Ning [12]. Other WSN security research works include authentication [13], denial-of-service (DOS) attacks [14], routing security [15], group security [16,17], multiple-key management [18,19], and simple system-level security analysis [20–23].

Why do those WSN security schemes not work well in WSAN environments? One of the common drawbacks of those sensor network security schemes is that they do not integrate security with a *hierarchical* sensor network *routing* architecture and specific characteristics of WSANs (such as the coexistence of actuators and sensors). Because the sensors may only want to report data to the nearby actuators, it will cause much security overhead if we build secure links between any two nodes. It is necessary to reduce key management overhead through cluster-based communication architecture around each actuator [2,3]. In this work, we integrate WSAN security issues with a proposed ripple-zone-based WSAN routing architecture (Section 8.3.4). We will show that a clustering-based re-keying scheme can save lots of energy compared to those works based on a general flat routing topology. (Note that energy consumption is the top concern in tiny, battery-driven sensors [3].)

As shown in Figure 8.2, most traditional sensor network security schemes just focus on end-to-end security issues (top diagram) and ignore WSAN routing details. They do not build security above the low-energy routing

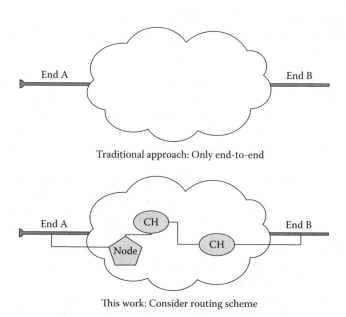

Traditional approach: Only end-to-end

This work: Consider routing scheme

Figure 8.2   (Top): separate security analysis from routing; (Bottom): integrate security and routing.

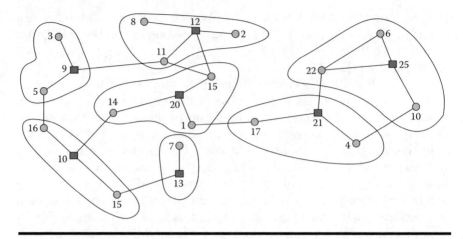

Figure 8.3 Pebblenets [32].

architecture and just simply assume that the entire network uses tree- or flat-based topology. Without concern for the routing details, it is impossible to prevent person-in-middle attacks or hop-to-hop security. Our scheme integrates security with a WSAN topology discovery procedure (Figure 8.2, bottom).

A related work close to ours is Pebblenets (Figure 8.3) [32]. It proposes a cluster-based routing architecture. However, it has several major shortcomings:

1. It only assumes a simple homogeneous architecture (i.e., only sensors) and thus does not apply in WSAN platforms that have both high-energy actuators and low-energy sensors in one network. Our scheme will use a two-level routing/security protocol to achieve key refreshing among high-level actuators and low-level sensors.

2. Another downfall to Pebblenets is the idea of having all keys generated by a backbone sensor node. Therefore, if a crucial backbone sensor is captured during key generation, new keys can be known, resulting in full system compromise. In our scheme, the BS and actuators together control new key generations. We use the concepts of Backbone Key and Session Key (Section 8.3.2) to securely update keying information.

3. We have also made other significant improvements, such as the concept of *ripple key*, that can achieve asynchronous broadcast authentication (i.e., without the synchronization assumption as in μTESLA [9]). (See Section 8.4 for details).

8.4 High-Level Security

8.4.1 Two-Level Routing/Security Architecture

Before discussing our WSAN security scheme, we first explain our proposed scalable WSAN routing architecture that serves as the basis of our security implementation. As the prerequisite of WSAN trustworthiness, we argue that it is very important to design a hierarchical, energy-efficient routing scheme compatible with the specific network characteristics of WSANs because both reliable packet transmission (internal trustworthiness) and security key management (external trustworthiness) need a low-energy routing protocol. We have created a Member Recognition Protocol (MRP) to allow actuators and sensors to self-organize themselves into separate *domains*, with each actuator as the domain center (see also [33,34] for details on the WSAN routing establishment scheme). After running our MRP, each actuator will be aware of its domain members.

As shown in Figure 8.4, within the domain of each actuator, we further propose the concept of a *ripple zone* (RZ) around each actuator, in which sensors are assigned to different *ripples* based on their distances, in number of hops, from their actuator. Some sensors are then chosen as *masters*, and called *cluster heads* (CHs), based on our Topology Discovery Algorithm (TDA) [33]. Each *master* aggregates data from the sensors in its zone before it transmits data to a *master* in the next *ripple* that is closer to the actuator, (i.e., with a smaller number of hops to the actuator).

In the remainder of this chapter, when we mention *node*, it can refer to an actuator, a CH, or a common sensor in a zone. We use CHs/masters, sink/base station (BS) alternatively. In Section 8.4.5, we further discuss the secure routing establishment procedure among sensors.

Due to sensor failure/addition, the above RZ-based routing architecture may break down. Thus, our above routing algorithm needs to run

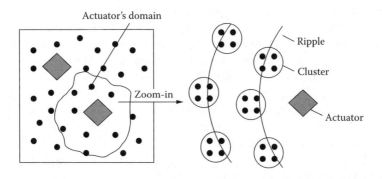

Figure 8.4 Proposed ripple-zone-based WSAN routing.

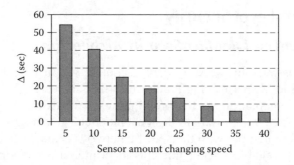

Figure 8.5 Re-routing period versus topology changing speed.

periodically. Obviously, the faster the topology changes, the shorter the routing update interval. Here let us determine the proper update interval value, T. We denote d as the sensor radio range. Suppose the total number of sensors is n and p is the probability of a sensor becoming a CH. Because at any time there are np zones, the average number of sensors in each zone, n', should be $n' = n/np = 1/p$. We denote P_{in} as the probability that a current sensor fails in its zone, and P_{out} as the probability that a new sensor is added to the current zone. Thus, the probability that no new sensor enters the current zone is: $P_{no\_in} = (1 - P_{in})^{n-n'}$, and the probability that no sensor fails in the current zone is: $P_{no\_out} = (1 - P_{out})^{n'}$. The probability that the zone changes topology is then as follows:

$$P_{update} = 1 - P_{no\_in} P_{no\_out} = 1 - (1 - P_{in})^{n-1/p}(1 - P_{out})^{1/p} \qquad (8.1)$$

The value of P_{in} and P_{out} can be determined by the conclusions from Ning et al. [38] and is approved to be the function of sensor amount changing speed and d. Given a target *topology change probability*, we can determine the topology update interval such that the practical topology change probability is below that target value. Suppose 500 sensors are randomly distributed in an area with a radius of 1 km and each sensor has a transmission range of 20 m. Also suppose that each zone has an average of 50 sensors, with average sensor amount changing speed as 5 sensors/s. If we want to make the topology change probability less than 10% (of the target topology change probability), we get the tree-zone updating interval of 5.5 s. Figure 8.5 shows the relationship of routing update interval Δ as a function of topology change speed.

Security assumptions: Before the discussion of our security scheme, we make the following reasonable assumptions, just as in most current sensor network security schemes:

1. The BS is always trustworthy and is located in a safe place without power or memory limitation.

2. Before node deployment (i.e., in the predeployment phase), all sensors/actuators share an initial *global key* with the base station (BS).

3. Each sensor or actuator also shares a 1:1 *initial pair-wise key* with the BS.

4. Each sensor has a unique ID with enough length (just like a unique credit card number). Thus, the BS can use the sensor ID to distinguish between legitimate IDs and illegitimate ones.

5. All the above keying or ID information is stored in a semi-permanent memory and will become invalid or be automatically self-erased after a pre-determined expiration timer. For example, if the network deployment time is 11 AM, we can set up the expiration timer to 11:05 AM (i.e., 5 minutes of valid time). We assume that the enemies need a certain time (at least a few minutes) to break the sensor and get to know the security information. Immediately after the sensors are dropped to some area, those sensors will use the initial global key, initial pair-wise key, and sensor ID to securely establish the above ripple-zone-based WSAN routing architecture. Once the routing architecture is established and recognized in all the actuators, sensors, and CHs, the re-keying protocol will be used to periodically refresh all types of keys (see Section 8.3.2 and Section 8.5 for more details).

When a new node joins in a WSAN, it will also use the above keying information (i.e., initial global key, initial pair-wise key, and sensor ID) to get recognized by the existing nodes.

8.4.2   High-Level Security (among Actuators)

Based on the above RZ-based routing architecture, we further propose a low-energy, low-complexity, and hierarchical (two-level-based: high-level and low-level) key management scheme.

In the high level[2] (among actuators), two types of keys exist: (1) a *session key* (SK) is used to secure the transmission of data packets; and (2) a *backbone key* (BK) is used to secure control packets that include SK re-keying information. Figure 8.6 shows the relationship between these two keys. SKs need to be re-keyed periodically to defeat attacks. However, the BK is refreshed in an event-triggered way (typical events include actuator insertion, death, or compromise). The WSAN base station (i.e., the sink) can use any well-known Group Communication Security protocol [35,36] to perform BK re-keying among actuators.

To ensure secure SK re-keying, the sink initially uses BK management schemes [35,36] to encrypt, authenticate, and transmit a control packet,

---

[2] Next section (i.e., Section 8.5) will discuss *low-level* security (among sensors belonging to the domain of an actuator).

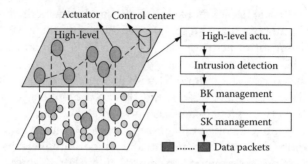

Figure 8.6 Backbone key (BK) and session key (SK) (Note: *High-level* security occurs among all actuators; while *low-level* security occurs among all sensors in an actuator domain.)

which includes an SK to be used in the current sensing data transmission. Note that control packets are different from data packets. The former is secured by BK management schemes, while the latter should be secured by SKs.

Every certain interval, a new SK will be distributed to all actuators. All new SKs are derived from a one-way hash function H as in Perrig et al. [9]. We do not use a MAC (message authenticated code) for SK authentication because receivers can use the one-way property of SK sequence to check whether the received SK belongs to the same sequence or not. We call such an SK authentication concept an *implicit authentication*, which works based on the following key chain scheme (Figure 8.7), which is also used in the LiSP scheme [37].

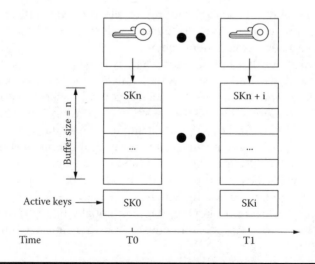

Figure 8.7 Key-chain scheme among actuators.

The sink first precomputes a long one-way sequence of keys: SK_M, $SK_{M-1}, \ldots, SK_n, SK_{n-1}, \ldots, SK_0$, size $M \gg n$, where $SK_i = H(SK_{i+1})$. Initially, only SK_n (instead of the whole M-size key sequence) is distributed to each actuator. Then an actuator can utilize H to figure out $SK_{n-1}, \ldots, SK_0$. The n keys $SK_n, SK_{n-1}, \ldots, SK_1$ are stored in a local key buffer with a length of n. However, SK_0 is not in the buffer because it is used for the current data packet encryption/decryption. After the initial SK_n delivery, the sink periodically sends $SK_{n+1}, SK_{n+2}, \ldots, SK_M$ (one key distribution in each period) to all actuators. To achieve communication confidentiality, each new key is encapsulated into a re-keying control packet that is encrypted with the BK or the currently active SK.

After receiving a new SK, the actuator keeps applying H(hash function) to it for some time, trying to find a key match in its key buffer. For example, assume that an actuator receives a new key SK_j. Also assume that its key buffer already holds n SKs as follows: $\{SK_i, SK_{i-1}, \ldots, SK_{i-n+1}\}$. If the actuator finds out that $H(H \ldots (H(SK_j))) \notin \{SK_i, SK_{i-1}, \ldots, SK_{i-n+1}\}$, (i.e., implicit authentication fails), SK_j will be discarded. Otherwise, if implicit authentication is successful, the key buffer is shifted one position. The SK shifted out of the buffer is pushed into the *active* key slot to be used as the current SK (see Figure 8.7), and the empty position is filled out with a new key SK'. SK' is derived from the received SK_j through H and meets the following two conditions: $SK' = H(H \ldots (H(SK_j)))$ and $H(SK') = SK_i$.

The reason for using a key buffer is to tolerate multiple key losses. As shown in Figure 8.8, even if up to $(n-1)$ SKs are lost due to unreliable wireless channels (such as radio shading, fading, and communication noise), as long as the nth SK received is authenticated, we can still restore the lost SKs in the key buffer by applying the one-way hash function. Using a key buffer, the lost SK will not influence the current security session until a maximum delay of $(n-1)$ re-keying periods. This feature also makes SK management robust to clock skews.

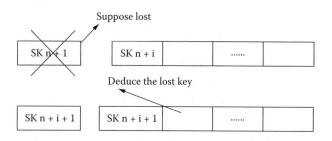

Figure 8.8   Handling key loss.

As discussed above, SK re-keying is conducted periodically to overcome active attacks (Figure 8.5). However, the SK should be reset immediately under any of the following conditions:

- One of the actuators is compromised by attackers.
- An actuator runs out of energy and thus stops operations.
- A new actuator is added.
- An actuator loses $n+1$ keys and thus cannot deduce any lost keys.
- The sink distributes all M keys and needs to build a new key sequence.

If the sink detects topology change (due to actuator addition/death) or detects node compromise through intrusion detection schemes, it immediately renews the backbone key to BK_{new} through group communication protocols [35] and generates a new key sequence:

$$SK'_M, SK'_{M-1}, \ldots, SK'_{n'}, SK'_{n'-1}, \ldots, SK'_0, (M \gg n'), SK'_i = H(SK'_{i+1})$$

where n' is the new key buffer size in each actuator[3]. The sink then sends out a new group of parameters to each actuator as follows:

$$Sink \rightarrow actuator : E_{BK_{NEW}}(\Delta' | n' | SK'_{n'}) \tag{8.2}$$

where Δ' is the new re-keying period.

On receiving these new parameters, an actuator will perform the following tasks:

1. Delete all old SKs in the previous buffer.
2. Rebuild a new buffer with length of n'.
3. Calculate the remaining n' keys $SK'_{n'-1}, \ldots, SK'_0$ based on $SK'_{n'}$ and the hash function H.
4. Choose SK'_0 as the current active session key for data encryption.
5. Invoke a new timer that expires every Δ'.

When timer expires, the sink sends out a new key based on the following rule:

$$Sink \rightarrow Actuator : E_{BK}(SK'_{n'+i+1}) \ at \ t = i * \Delta' \tag{8.3}$$

If the sink has distributed the whole key sequence (after $n \times \Delta'$), it will also send out a new set of parameters as shown in Formula 8.2. If an actuator misses n keys, the sink must send a message with $E_{BK}(\Delta' | n' | SK'_C)$ to that actuator, where SK'_C is the current session key.

---

[3] Each time, we require the actuators to create buffers with a new size n' to make attacks more difficult.

We can achieve better security by using a shorter re-keying period Δ' (i.e., more frequent key updating). However, this will lead to greater communication overhead and increased hash calculation. There also exist other trade-offs. For example, the larger the buffer size n, the more tolerant to key loss an actuator is, which is because it can recover more missed keys based on a one-way hash function. However, a large buffer requires more memory and calculation overhead.

Figure 8.9 shows the procedure of the security protocol in the high-level nodes (actuators). It is implemented in our simulations (Section 8.5) and experiments (Section 8.7).

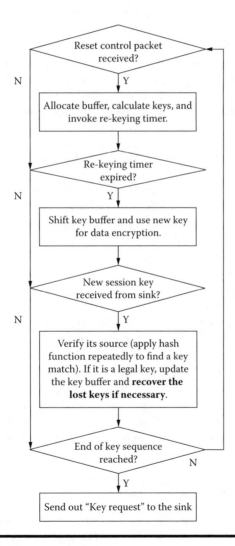

Figure 8.9   High-level security protocol.

8.5 Low-Level Security

8.5.1 Low-Level (in the Domain of Each Actuator) Re-keying

A unique issue in WSAN security is that the selection of key-sharing schemes should consider the impact on in-networking processing [9]. For example, data aggregation is necessary for reducing communication overhead from redundant sensed data. If one simply adopts a key scheme like pair-wise key (that is shared between only two nodes), memory limitations will prohibit a *master* from maintaining all the keys necessary to aggregate data from its zone sensors. In addition, building an end-to-end secure channel between any two nodes is inadvisable, because intermediate sensors/actuators may need to decrypt and authenticate the data collected from multiple sensors. Because different types of messages exchanged among sensors have different security requirements (such as data aggregation security), a single keying mechanism may not be suitable for all cases. Thus, multiple keys should be introduced in *low-level sensors in an actuator domain*. We have defined multiple types of keys for different security purposes as follows:

1. *Master-to-Actuator Key (MAK):* An MAK is shared between each master and its domain actuator. It is used for direct master-to-actuator secure communication. An MAK is generated based on a high-level session key (SK) as follows: MAK = f_{SK} (Master-ID).
2. *Inter-Master Pairwise Key (MPK):* Occasionally, secure channels need to be established between two masters that belong to two actuator domains.
3. *Sensor-to-master Pair-wise Key (SPK):* A sensor-to-master pair-wise key is shared between a master and each of the sensors in its zone.
4. *Zone Key (ZK)*, also called Cluster Key (CK): Zone keys are used for data aggregation and also for the propagation of a query message to the whole cluster. Each ZK is shared among all sensors in the same cluster.
5. *Ripple Key (RK):* A ripple key is used to achieve hop-to-hop security in an actuator domain. To save security overhead, our scheme generates a new key based on a family of pseudo-random functions (PRF) $F(x)$ as follows: $K'_{K,x} = f_K(x)$ (K is last key and x is a random number). Section 8.4.4 provides a detailed procedure for using ripple keys.

8.5.2 Robustness to Key Losses

In High Level, we use the one-way property of hash functions to recover the lost keys between two re-keying events. In Low Level, because we do not use a one-way key chain (because it is not suitable for sensors that

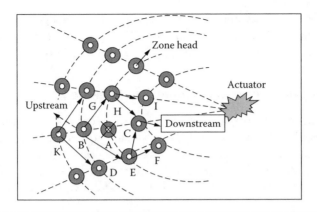

Figure 8.10   Ripple-to-ripple key loss recovery.

have less memory and computational capability than actuators), we recover lost *control packets* that include key information through a ripple-to-ripple retransmission scheme and a *local path repair* routing protocol [34]. A simple example demonstrating our key loss recovery scheme is shown in Figure 8.10, after master B sends out a control packet to A (the closest master to B), B sets up a timer $T1$. If $T1$ expires and B has still not received a positive acknowledgment packet from A, B will retransmit the packet to A. If B attempts more than three retransmissions without success, it regards A as unreachable (possibly due to an unreliable wireless link), and our scheme then finds another good path as follows: B checks the reachability of A's neighboring masters H and E. If either of them is reachable, B will transmit the packet to one of them. If both H and E are unreachable, our protocol goes back one ripple (i.e., to master K). K discovers B's neighboring masters and tries to deliver the packet. Eventually, a new ripple-to-ripple good path will be discovered. We will verify its efficiency through simulation experiments in Section 8.6.

8.5.3   Relationship between SPKs and ZKs

ZKs are very important for securing data aggregation in a local sensing area because they enable the whole zone to encrypt and decrypt data packets without the use of multiple keys. Note that ZK re-keying occurs after SPK re-keying. The basic procedure is as follows. Whenever a master M regenerates a new random key $ZK' = f_{ZK}(x)$ (where ZK is the last zone key and x is a random number), it sends ZK to each of its zone members using the corresponding SPKs as follows:

$M \rightarrow actuator$:

$$Rekeying\_No | Master\_ID | E_{SPK(i)}(ZK' | Rekeying\_No | Master\_ID) \quad (8.4)$$

Each zone member analyzes and decrypts this message, then stores the new ZK in its buffer. The re-keying of ZK is performed periodically (the control timer is the same as *Level* Δ; see Section 8.3.1).

8.5.4 Application of the Ripple Key Scheme

The aforementioned ripple key scheme can overcome the time synchronization problem [9] when used for broadcast authentication; that is, each zone ensures that a broadcasting message does come from its own actuator versus other sources. To disseminate a message, an actuator broadcasts a separate message (called *Cmd_MSG*) for different ripples as follows:

$$Actuator \rightarrow Ripple[i]:$$

$$MSG|Actuator - ID|Tran - ID|$$

$$MAC_{RK(i)}(MSG|Actuator - ID|Tran - ID) \qquad (8.5)$$

where *Actuator − ID* helps each master distinguish its own actuator from others. *Tran−ID* helps avoid routing loops as follows: a master identifies different commands through Transaction ID (i.e., *Tran−ID*) and discards duplicate ones. MAC (*message authenticated code*) is derived from the RK of ripple level *i*. Intermediate masters that do not belong to ripple *i* cannot tamper with the above message without detection because they do not know RK *i*. After each master receives an authenticated message, it will use the *zone key* to broadcast to all sensors in its zone.

If a message arrives at the masters in the same ripple with variable delays, there could exist the following attack: a master A in ripple *i* that receives a message MSG earlier than other masters in the same ripple can fake a new message MSG and generate a new MAC using its RK *i*, and then send it to another master B in ripple *i*. If B has not received MSG yet, it will be unable to identify the forged message MSG. To protect the system from this attack, we use our aforementioned *Inter-Master Pair-wise key* (MPK) as follows:

1. We require that each master in ripple *i* maintain a few MPKs in its cache (called ripple-MPKs) with its close upstream and downstream masters that are located in ripple *i − 1* (upstream) and *i + 1* (downstream), respectively. We use *received signal strength* (RSS) to determine the geographically close masters according to the following relationship between the RSS and the distance x from the transmitter [5]:

$$RSS_{dB} = -10\gamma \times Log(x) \qquad (8.6)$$

where γ is the propagation path-loss coefficient [5].

2. Each time a master in ripple $i - 1$ receives a message and finds out that it is supposed to be received by its next ripple, it will encrypt it by the corresponding downstream ripple-MPKs and relay the following message to the downstream masters (in ripple i):

$$Message - to - downstream(i) : E(Ripple\_MPK, Cmd\_MSG)$$
$$(8.7)$$

3. When a master receives the above message, it will try each of its upstream ripple-MPKs to recover the contents of the original message (i.e, Cmd_MSG).

The above scheme can prevent message spoofing from masters in the same ripple because a master only receives and decrypts messages from its upstream masters.

Because the message encrypted by ripple keys (RK) are transmitted in a hop-by-hop way, nodes on the smallest circle (i.e., closest to the actuator) will take a large relaying load and will die more quickly than others. This is a common problem in the uplink (sensors-to-actuator/sink) transmission of any sensor network routing protocol. We will include this issue as part of our future work. However, in the downlink (actuator-to-sensors), we can simply use the broadcast scheme because an actuator has enough antenna power to cover its entire domain.

8.5.5 Detailed Low-Level Security Procedure

In our security protocol, we use three types of messages: *setup messages*, *data messages*, and *system messages* for different functions. *Setup messages* are used for tasks that pertain to setting up the system, such as node/CH addition, authentication, and key refreshing. *Data messages* are used for sending and receiving data, as well as generating data queries. *System messages* are generally used for tasks that affect the system or its topology (e.g., node removal). A breakdown of all the messages within our security protocol can be seen in Figure 8.11. Every message sent passes through the routing and security layer, where it gets a routing header that includes (1) message source, (2) message destination, (3) number of hops, and (4) list of hop IDs.

As an example, we briefly mention two messages to be used in our secure routing discovery protocol (more details will be provided later):

1. *RREQ message:* When a node needs to communicate with a node for which it does not have an entry in its routing table, that node sends a Route Request message (RREQ) to all of its neighbors. The neighbors then forward the RREQ to their neighbors, each appending their node ID. This process continues until the RREQ is received by the destination, which replies to the first RREQ message received.

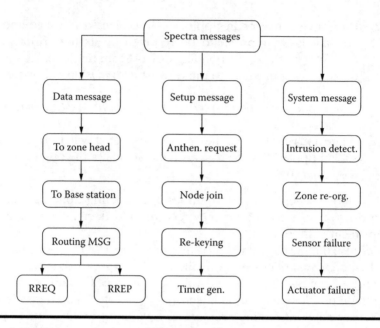

Figure 8.11 Messages exchanged between nodes.

2. *RREP message:* Route Reply messages (RREP) are generated in response to an RREQ, by the destination node, or a node that knows a route to the destination. An RREP contains the complete route from source to destination. This message then follows the path that it specifies back to the requesting node (the source). Unlike RREQs, RREP messages are eavesdropped by all nodes that receive them and the route in the RREP is extracted to fill in the intervening nodes routing table. This eavesdropping takes place to reduce unnecessary routing communication.

Recall that in Section 8.3.1, we made five basic assumptions. Based on that initial keying information (such as *initial global key, initial pairwise key, and sensor ID*), we can securely establish our ripple-zone-based routing architecture (see Figure 8.4 in Section 8.3) as follows.

This phase of the protocol is responsible for setting up the communication routes for inter-cluster and intra-cluster routing. A diagram depicting an overview of the route establishment phase is prented in Figure 8.12.

We have created a clustering algorithm [33] to select some sensors in an actuator domain to become cluster heads (CHs). Once a CH is authenticated using an initial pair-wise key by the base station, it broadcasts an advertisement, encrypted with the *initial global key*. Advertisements contain the cluster ID number and the assigned *cluster key* by the base station.

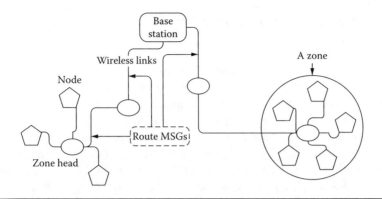

Figure 8.12   Overview of the secure route establishment phase.

Nodes listen to these advertisements and record their RSSs. The strongest recorded RSS is associated with the nearest CH, and the node sends a *cluster join* message to this CH, encrypted with the cluster key. The cluster key is received through the cluster advertisement message (Figure 8.11).

The BS keeps track of network topology through the cluster member registry of each CH. Whenever there is a change in the topology of a cluster, a new *cluster organization report* is sent to the BS. This knowledge is used in the event of CH compromise in order to reorganize the cluster.

After clusters are organized based on our clustering algorithm, but before the CH sends its first cluster organization report, CHs find a route to the BS. If the BS is not one of its neighbors, then the CH broadcasts a Route Request (RREQ) message. A neighbor is defined as a node whose *Received Signal Strength* (RSS) is above a certain threshold, and every hop of route must occur between neighbors.

If the current recipient is not a neighbor of the requested destination, then it forwards the RREQ to all of its neighbors through a broadcast encrypted with the system key. It appends its own node ID to the route contained within the RREQ before forwarding the message. In the event that the RREQ is intended for one of the neighbors of the current recipient, the modified RREQ is forwarded only to the destination.

After the above cluster-based routing architecture is established in each actuator's domain, then at a certain time (the timer is determined by the actuator mobility and security requirements; for example, a shorter timer can make the security performance higher), the actuator will broadcast a new global key to its entire domain encrypted by its BK (see Section 8.3.2). It will also broadcast new Master-to-Actuator (pair-wise) keys encrypted by the corresponding *last* Master-to-Actuator keys to each CH in order to refresh the secure communication keys between each CH and the actuator.

8.5.6 Dynamic Security Scheme

8.5.6.1 Sensor Addition

Sensors trying to add themselves to an existing WSAN must undergo authentication using their *initial global key, initial pair-wise key, and sensor ID* (refer to Section 8.3.1 security assumptions). The BS is, in fact, preloaded with such information before a new sensor joins. Once authenticated, the node broadcasts a *cluster joining* message, encrypted with the initial global key. CHs reply to the request with their cluster ID and cluster key, encrypted with the latest BK. The BS then helps forward the CHs' response to the new sensor using the initial pair-wise key with the sensor. The sensor keeps track of the RSS of all replies and joins the cluster with the highest RSS (i.e., the closest cluster). The node ID of the new sensor is then added to the *cluster member registry* of the CH that is eventually forwarded to the BS (in the cluster organization report phase).

8.5.6.2 Sensor Death

When a node's available power drops below a certain threshold, that node sends a *Node Death* message to its CH. The CH then sets a timer to three times the predicted remaining lifetime of the node. Once this timer expires, the CH removes this node from its *cluster member registry* and broadcasts a notification to its cluster. This message instructs all nodes in the cluster to eliminate the dying node from their routing tables.

8.5.6.3 New CH Identification

A common sensor can request to become a CH if its CH is close to being out of power. Once a joining CH has been authenticated by the BS through the latest Zone key and Master-to-Actuator key, it proceeds to set up a cluster. The CH broadcasts a Cluster-Advertisement message that is encrypted with the last SK (Session Key, see Section 8.3.2). Any node that receives a stronger signal from the new CH compared to the signal from its current CH replies with a *cluster joining* message, encrypted with the last SK. This message is broadcast and contains the node ID and the CH ID. This enables the node's current CH to remove the node from its cluster member registry, and the new CH adds that node to its registry. Eventually, the BS is notified of this change through a *cluster organization report*.

8.5.6.4 CH Death

CH death is handled in a fashion similar to node death. When a CH reaches a certain power level, it sends a message to the BS notifying it of its state. The BS sets a timer for three times the predicted remaining lifetime of the CH. When the timer expires, the BS uses a broadcast to notify the corresponding cluster that its CH has died. This causes the nodes in the cluster to reorganize themselves and appoint the node with the largest remaining power to be the new CH.

8.6 Performance Analysis

8.6.1 Jist+SWANS-based WSAN Security Simulation

The WSAN security performance analysis results have been obtained through a Java-based simulation engine. (Section 8.7 further describes our hardware test results). The simulation engine is comprised of JiST [39] (Java in Simulation Time), and SWANS [40] (Scalable Wireless Ad Hoc Networks Simulator). JiST was created to simulate time in Java, while SWANS was created to simulate WSANs.

To reduce keying information loss, in addition to the aforementioned local loss recovery scheme (Section 8.2), we have also implemented a reliable Transport Layer protocol above our cluster-based two-level routing protocol (see Figure 8.13). We implement a CH-by-CH NACK (Negative ACKnowledge) based reliability scheme. A gap in the sequence number of sent packets indicates packet loss. Each CH maintains a list of missing packets. When a loss is detected, a tuple containing a source ID and sequence number of the lost packet is inserted into this list. Entries in the *missing packets* list are piggybacked in outgoing transmissions, and CHs infer losses by overhearing this transmission. CHs keep a small cache of recently transmitted packets, from which a child can repair losses reported by its last hop CH. In addition to CH-by-CH loss recovery, we also implement end-to-end recovery. This is because heavy packet losses can lead to large missing packet lists that might exceed the memory of the motes.

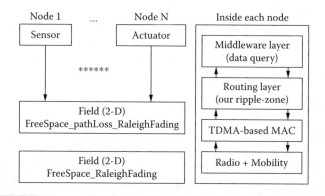

Figure 8.13 WSAN simulator software architecture.

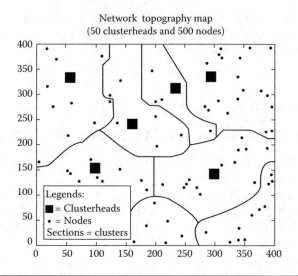

Figure 8.14   WSAN simulation topology.

Our end-to-end recovery scheme leverages the fact that the base station has significantly more memory and can keep track of all missing packets. The base station attempts CH-by-CH recovery of a missing packet. When one of the CHs notices that it has seen a packet from the corresponding source, but does not have a cached copy of that packet, it adds that recovery request to its missing packets list. This request is propagated downward in this manner (using the same mechanisms described for CH-by-CH recovery) until it reaches the source. Because the source maintains generated packets in its memory, it can repair the missing packet. Figure 8.13 also shows that our WSAN simulator considers the actuator mobility, which causes topology change and has a direct impact on the setup of re-keying period (Section 8.3.1).

Figure 8.14 shows our generated *ripple-zone (RZ)* (see Section 8.3.1) routing topology. Our simulator can simulate a large-scale WSAN with over 500 nodes and produce simulation results within less than one hour, which is faster than OPNET [42].

8.6.1.1   Energy Model

SWANS does not natively have a function or layer that tracks energy consumption, which is one of the most important metrics in WSANs. A battery layer was then added to keep track of energy consumption during simulation runtime. We use the same radio model as discussed in Heinzelman et al. [4] which is the first-order radio model. In this model, a radio dissipates $E_{elec} = 50$ nJ/bit to run the transmitter or receiver circuitry and

$amp = 100$ pJ/bit/m^2 for the transmitter amplifier. The radios have power control and can expend the minimum required energy to reach the intended recipients. The radios can be turned off to avoid receiving unintended transmissions. An r^2 energy loss is used due to channel transmission [5]. The equations used to calculate transmission costs and receiving costs for a *k*-bit message and a distance *d* are:

Transmitting:

$$E_{Tx}(k, d) = E_{Txelec}(k) + E_{Txamp}(k, d)$$
$$E_{Tx}(k, d) = E_{elec} * k +_{amp} * k * d^2 \tag{8.8}$$

Receiving:

$$E_{Rx}(k) = E_{Rxelec}(k) \quad E_{Rx}(k) = E_{elec} * k \tag{8.9}$$

Receiving is also a high-cost operation; therefore, the number of receives and transmissions should be minimal. In our simulations, we used a packet length *k* of 2000 bits. With these radio parameters, when d^2 is 500 m^2, the energy spent in the amplifier part equals the energy spent in the electronics part and therefore the cost to transmit a packet will be twice the cost to receive.

8.6.2 Performance Test Results

Because most WSAN nodes are tiny sensors, the security protocols should have low communication energy consumption. We have investigated the energy efficiency of our RZ-routing-based (see Section 8.3) security scheme. Figure 8.15 clearly shows that our scheme has the lowest energy consumption (within all nodes within a unit time) compared to the security based on other routing schemes such as Pebblenet (that is based on a simple one-level cluster scheme) [32] and the security scheme based on flat-topology routing strategy (we used the scheme in [8]).

Figure 8.16 shows that (before protocol optimization) a sensor that is selected as CH (cluster head) can lose much more energy than common sensors in its cluster. This is due to the heavy traffic load handled by a CH, such as data aggregation results from all cluster sensors, inter-cluster communications (between two CHs), cluster key management, etc. Therefore, our RZ-based routing/security protocol adopts a CH *role rotation* strategy; that is, every certain time, another cluster member will replace the role of the current CH. As shown in Figure 8.17, we randomly picked up three clusters and measured the sum of the remaining energy in all nodes of each cluster. All clusters have similar energy consumption amount.

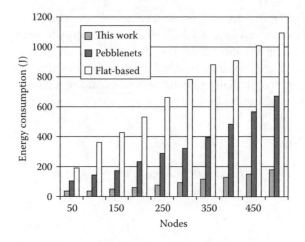

Figure 8.15 Sum of energy consumption in all nodes.

In each actuators domain, the selection of the number of clusters has a direct impact on the system energy consumption. As shown in Figure 8.18, there could exist an optimal value of the number of clusters. This can be explained as follows. When we form more clusters, more energy consumption comes from the inter-cluster communication and high-level key management (BKs and SKs). On the other hand, when forming fewer clusters, the multi-hop communication overhead in each cluster (i.e., intra-cluster routing) becomes intolerable.

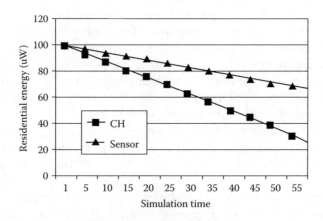

Figure 8.16 Energy consumption in CHs and common sensors.

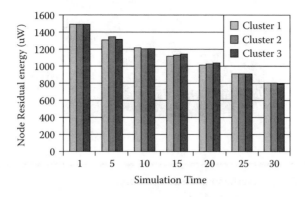

Figure 8.17   Use CH role rotation to achieve load balance.

Based on the above-mentioned energy model (Section 8.5.1), we have collected the statistical results on the energy consumption of different operations in our WSAN security protocol (see Figure 8.19). Most of the energy (86 precent) is consumed in data transmission. The local procession (such as hash function calculations) uses much less energy (only 4 percent) than wireless communications (86 percent), which implies the importance of reducing security/routing overhead in WSANs. As shown in Figure 8.19, our WSAN scheme has low complexity due to its two-level ripple-zone topology management and symmetric cryptography protocol (the routing protocol overhead is 7 percent and the security key management overhead is only 3 percent).

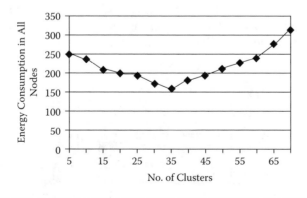

Figure 8.18   Balance of the number of clusters.

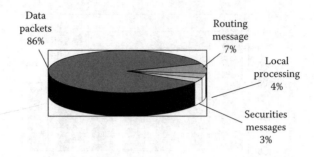

Figure 8.19 Energy overhead of each operation.

We further investigate the effect of encryption key size on power consumption. As can be seen in Figure 8.20, the effect of key size on power consumption of all nodes fits an exponential distribution. Increases in the level of key size can double the power consumption required for encrypting a packet. However, as shown in Figure 8.19, the amount of power consumed for encryption (3 percent) is fairly minimal compared to the overall battery capacity. Therefore, although higher levels of encryption require far more power, the power consumed by encryption (3 percent) is far less than the power consumed by communication overhead (86 percent + 7 percent = 93 percent).

8.6.2.1 Reliability Test

We adopt both a ripple-to-ripple and end-to-end loss recovery scheme to handle packet losses (see Section 8.4.2 and Section 8.5.1). Figure 8.21 shows that other security schemes, such as Pebblenet [32] that uses only end-to-end recovery or one-level security scheme [8] that uses a simple flooding-based recovery, have a much higher keying packet loss rate than our security scheme that is based on RZ architecture (Section 8.3.1).

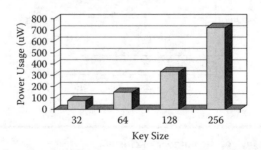

Figure 8.20 The impact of key size on energy overhead.

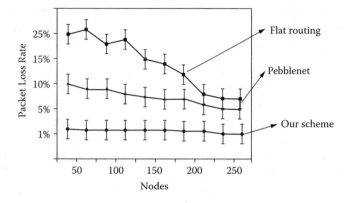

Figure 8.21   Reliability test.

8.7   Security Analysis

8.7.1   *Protection from Various WSAN Attacks*

Now we analyze the protection of a WSAN through our security scheme from various attacks.

8.7.1.1   *BK Attacks among Actuators*

Because the distribution of new SKs depends on control packets encrypted by BK that is managed by group security schemes, it is possible for an attacker to compromise the current BK and thus can attack any future SK disclosures. Our scheme can minimize the impact of this attack through our *buffered key chain* scheme. Thanks to the SK buffer, there is a delay between the receiving of new SK and the actual applying of this new SK. Suppose the distribution interval is Δ (i.e., the re-keying period) and n is the buffer length. Then until $n \times \Delta$ later we may use this new SK. As long as we can detect the BK compromise within $n \times \Delta$ and renew BK and SKs through a new High Level *initialization procedure* that can help each actuator rebuild a new SK buffer chain, the data packets will maintain normal security performance.

8.7.1.2   *SK Attacks among Sensors in Each Actuator Domain*

The attacker may modify the transmitting SK, inject phony SK, or use wireless channel interference to damage control packets. These attacks may also result in data replay and denial-of-service (DoS) attacks. Our scheme can easily defeat these attacks through *implicit authentication*. Thanks to the one-way characteristics of the hash function keys, any false SKs cannot pass the *authentication test*; that is, after L times ($L \leq n$) of using hash function, if we still cannot satisfy the following formula, we will regard that it is a false SK: (in the following formula, SK_{FAKE} is a false SK and SK_{NOW} is the SK currently used.)

$$\underbrace{H(H(\dots(H(SK_{FAKE})\dots)))}_{L} = SK_{NOW} \qquad (8.10)$$

We have also included a rule in our security protocol: whenever a sensor continuously fails the *authentication test* for three times, it will send back an attack-detection message to its actuator to trigger a new High-Level *initialization procedure*.

8.7.1.3 Relay Attacks

Such attacks cannot succeed because a sensor ensures that a new received SK will not only pass the *authentication test*, but also be unique.

8.7.1.4 Periodical Attacks

Possibly an attacker can issue an attack at a fixed time if he or she knows the value of Δ' (i.e., the re-keying period). We address this problem through the addition of a random time interval $\epsilon(t)$ to Δ'; that is, the new re-keying period is $\Delta' + \epsilon(t)$, where $\epsilon(t) \leq \Delta'/8$.

8.7.1.5 Man-in-the-Middle Attacks

Our scheme can also defeat man-in-the-middle attacks (where an attacker fools the High-Level nodes as if he or she were a legal BK actuator) through a transmission of MAC in the High-Level *initialization procedure* as follows:

$$Sink \rightarrow Actuator : E_{BK_{NEW}}(\Delta'|n|SK_0|MAC(\Delta'|n|SK_0) \qquad (8.11)$$

8.7.1.6 Data-Level Attacks

Our scheme defeats it through SK re-keying every Δ', and inclusion of *Sensor_ID* and per-packet IV (initialization vector, which will also be updated from packet to packet) in the generation of keystreams to counter the keystream-reuse problem.

8.7.2 Security Overhead Analysis

We have used a first-order Markov chain model to analyze the calculation/communication overhead when incorporating our security features into actuator communications. As calculations involving the one-way hash

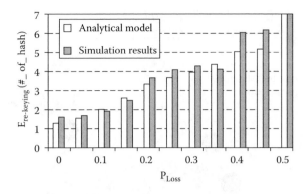

Figure 8.22   Security overhead.

function in local sensor processing consume the most energy [9], we therefore focus on the cost of computing hash functions during each re-keying
session. An actuator may fail to receive a new session key, or it may receive
an incorrect session key that cannot be authenticated using the hash function. Incorrect session keys may come from opponents attempting denial-
of-service attacks.

If the key chain buffer length is n, the probability of key loss is P_{Loss} and
the probability of key corruption is $P_{Corruption}$. We derive the expected times
for hash function calculations in a re-keying cycle, $E_{re\text{-}keying}[No\text{-}of\text{-}hash]$, as
follows [43]:

$$E_{Re\text{-}keying}[No\text{-}of\text{-}hash] =$$

$$\frac{2.5 - P_0}{2 - P_0^n} \left\{ n \bullet P_0^{n-1} + (1 - P_{failure}) \bullet \Sigma_n^{i=1}(i \bullet P_0^i) \right\} \qquad (8.12)$$

where $P_0 = P_{Loss} + P_{Corruption}$

Assuming $P_{Corruption} = 0.25$, we compare the simulation and analytical
results of $E_{re\text{-}keying}[No\text{-}of\text{-}hash]$ while varying P_{Loss} from 0.0 to 0.5. Figure 8.22
indicates the validity of our analytical model [43].

8.8   Hardware Experiments

We have used Crossbow sensor motes [41] to build our WSAN hardware
platform and have carried out a series of experiments to verify the efficiency of our proposed RZ routing-based security scheme. A WSAN node
includes two parts [41]: (1) microprocessor plus radio board (for sensor local

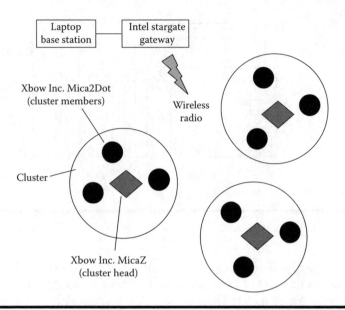

Figure 8.23 (Top): MicaZ node that serves as actuator; (bottom): hardware setup.

processing, and wireless transmissions), also called mote; and (2) sensor board (for detecting light, temperature, humidity, sound, and other types of data). Typically, these sensor motes have extremely low power (a few tens of milliwatts versus tens of watts for a typical laptop computer). When operating at 2 percent duty cycle (between active and sleep modes), we can achieve a lifetime of about six months (on a pair of AA batteries). We use MicaZ motes [41] that have IEEE 802.15.4 Physical/MAC layers modules. A Crossbow mote has a 4-MHz, 8-bit Atmel microprocessor. To build a WSAN prototype, we used a few MicaZ motes (Figure 8.23, top) as the high-energy actuators, each of which controls a cluster-based routing management in its domain (Figure 8.23, bottom). In each cluster, the CH and common sensors are implemented by Mica2Dot motes [41] that have less energy storage than MicaZ motes [41]. We have also implemented peer-to-peer routing among actuators and cluster-based routing around each actuator's domain.

8.8.1 Security Implementation

Our security design is similar to LiSP [37]. A stream cipher RC4 has been used to implement encryption/decryption between two MicaZ nodes because the stream cipher has a lower complexity compared to a block cipher. To address the keystream reuse problem, a sender includes its own sensor_ID into the generated keystream. For each message sent, the sender increments its own per-packet initialization vector (IV) by 1. Keystream

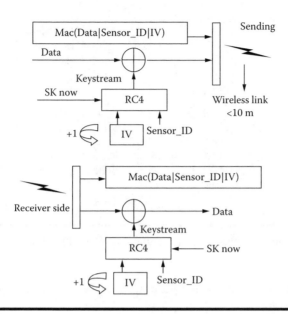

Figure 8.24 Cryptographic procedure.

uniqueness can therefore be ensured. The cryptographic procedure will follow the function components as shown in Figure 8.24. The Message Authentication Code (we denote it as MAC in Figure 8.24) is included for authentication purposes. To generate multiple keys, the pseudo-random functions (PRF) $f(x)$ are adopted to derive new secret keys based on the current session key SK_{NOW} and a random number x as follows: $KEY_{NEW} = f(SK_{NOW}, x)$, the generation of an x is based on the counter approach in Perrig et al. [9].

We measure the WSAN lifetime (the time until the first sensor failure occurs due to running out of battery) in the following three scenarios:

1. Using the suggested scheme, that is, the wireless communications occur in two levels:
   a. High-level among MicaZ motes (actuator-actuator): an actuator does not talk with a sensor belonging to another actuator's domain.
   b. Low-level: including inter-cluster and one-hop intra-cluster communications
2. Flat routing topology (i.e., all sensors organize into an ad hoc network without clustering).
3. Pebblenet [32]: all sensors organize into one-level cluster architecture without the backbone consisting of actuators. We let all sensors continuously work at a high sensing frequency (50 percent duty cycle).

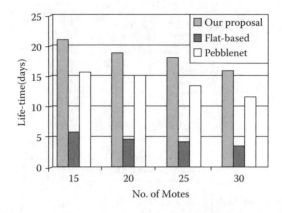

Figure 8.25   The impact of security protocol on WSAN lifetime.

Figure 8.25 shows that our proposal brings over four times longer lifetime than flat topology.

8.9   Conclusions

The focus of this work is the security design in an important information infrastructure: large-scale and low-energy wireless sensor and actuator networks (WSANs). The proposed schemes attempt to ensure that data is transmitted among actuators and sensors with the desired security (i.e., overcoming external network attacks). This work has proposed to seamlessly integrate WSAN security with a promising routing architecture that is scalable and energy efficient. To protect from active attacks in sensor networks, we have developed two-level re-keying schemes that can not only adapt to a dynamic network topology, but also securely update keys for each data transmission session. Moreover, to provide the security for the in-networking processing such as data aggregation in WSANs, we have defined a multiple-key management scheme in conjunction with the proposed *ripple-zone (RZ) routing* architecture.

References

1. C.-F. Huang, H.-W. Lee, and Y.-C. Tseng, A two-tier heterogeneous mobile ad hoc network architecture and its load-balance routing problem, *Mobile Networks and Applications*, Volume 9 , Issue 4 (August 2004) pages: 379–391.
2. I.F. Akyildiz and I.H. Kasimoglu, Wireless Sensor and Actor Networks: Research Challenges, *Ad Hoc Networks Journal*, (to appear), 2004.

3. I.F. Akyildiz, W. Su, Y. Sankarasubramaniam, and E. Cayirci, A Survey on Sensor Networks, *IEEE Communications Magazine*, August 2002.

4. W. Heinzelman, A. Chandrakasan, and H. Balakrishnan, Energy-Efficient Communication Protocol for Wireless Microsensor Networks, in *Proceedings of the Hawaii Conference on System Sciences*, January 2000.

5. T.S. Rappaport, *Wireless Communications*, Prentice-Hall, 1996.

6. D. Liu and P. Ning, Establishing Pairwise Keys in Distributed Sensor Networks, *The 10th ACM Conference on Computer and Communications Security (CCS'03)*, Washington D.C., October 2003.

7. D. Liu and P. Ning, Location-Based Pairwise Key Establishments for Relatively Static Sensor Networks, *2003 ACM Workshop on Security of Ad Hoc and Sensor Networks (SASN'03)*, October 31, 2003, George W. Johnson Center at George Mason University, Fairfax, VA.

8. W. Du, J. Deng, Y.S. Han, S. Chen, and P. Varshney, A Key Management Scheme for Wireless Sensor Networks Using Deployment Knowledge, in *IEEE INFOCOM'04*, March 7–11, 2004, Hong Kong.

9. A. Perrig, R. Szewczyk, V. Wen, D. Culler, and J.D. Tygar, SPINS: Security Protocols for Sensor Networks, in *Proc. of Seventh Annual International Conference on Mobile Computing and Networks (MOBICOM 2001)*, July 2001.

10. L. Eschenauer and V.D. Gligor, A Key-Management Scheme for Distributed Sensor Networks, in *Proc. of the 9th ACM Conference on Computer and Communications Security 2002*, Washington, D.C.

11. H. Chan, A. Perrig, and D. Song, Random Key Predistribution Schemes for Sensor Networks, *IEEE Symposium on Research in Security and Privacy*, 2003.

12. D. Liu and P. Ning, Establishing Pairwise Keys in Distributed Sensor Networks, *The 10th ACM Conference on Computer and Communications Security (CCS'03)*, Washington D.C., 2003.

13. S. Zhu, S. Setia, S. Jajodia, and P. Ning, An Interleaved Hop-by-Hop Authentication Scheme for Filtering of Injected False Data in Sensor Networks, *Proceedings of IEEE Symposium on Security and Privacy (S and P'04)*, Oakland, CA, May 2004.

14. A.D. Wood and J.A. Stankovic, Denial of Service in Sensor Networks, *IEEE Computer*, 35(10):54–62, 2002.

15. C. Karlof and D. Wagner, Secure Routing in Wireless Sensor Networks: Attacks and Countermeasures, *First IEEE International Workshop on Sensor Network Protocols and Applications*, May 2003.

16. G. Wang, W. Zhang, G. Cao, and T. La Porta, On Supporting Distributed Collaboration in Sensor Networks, *MILCOM 2003*, October 2003.

17. J. Zachary, A Decentralized Approach to Secure Group Membership Testing in Distributed Sensor Networks, *MILCOM 2003*, October 2003.

18. S. Zhu, S. Setia, and S. Jajodia, LEAP: Efficient Security Mechanisms for Large-Scale Distributed Sensor Networks, *The 10th ACM Conference on Computer and Communications Security (CCS'03)*, Washington D.C., October 2003.

19. F. Hu and N.K. Sharma, Security Considerations in Wireless Sensor Networks, *Ad Hoc Networks Journal, (accepted for publication)*, 2005.

20. Y.W. Law, S. Etalle, and P.H. Hartel, Key Management with Group-Wise Pre-Deployed Keying and Secret Sharing Pre-Deployed Keying, *EYES Project (Europe)*; available through GOOGLE search engine.

21. S. Slijepcevic, M. Potkonjak, V. Tsiatsis, S. Zimbeck, and M.B. Srivastava, On Communication Security in Wireless Ad-Hoc Sensor Network, *Eleventh IEEE International Workshops on Enabling Technologies: Infrastructure for Collaborative Enterprises (WETICE'02)*, June 10–12, 2002, Pittsburgh, PA.

22. M. Chen, W. Cui, V. Wen, and A. Woo, Security and Deployment Issues in a Sensor Network; downloadable from http://citeseer.nj.nec.com/chen00security.html

23. Y.W. Law, S. Dulman, S. Etalle, and P. Havinga, Assessing Security-Critical Energy-Efficient Sensor Networks, Department of Computer Science, University of Twente, Technical Report TR-CTIT-02-18, July 2002.

24. T.W. Carley, M.A. Ba, R. Barua, and D.B. Stewart, Contention-Free Periodic Message Scheduler Medium Access Control in Wireless Sensor/Actuator Networks, in *Proc. of Real-Time Systems Symposium*, Cancun, Mexico, December 2003.

25. M. Chow and Y. Tipsuwan, Network-Based Control Systems, in *Proc. of IEEE IECon 2001 Tutorial*, Denver, CO, pp. 1593–1602, November 28–December 2, 2001.

26. B.P. Gerkey and M.J. Mataric, A Market-Based Formulation of Sensor and Actuator Network Coordination, in *Proc. of the AAAI Spring Symposium on Intelligent Embedded and Distributed Systems*, Palo Alto, CA, March 25–27, 2002, pp. 21–26.

27. M. Haenggi, Mobile Sensor-Actuator Networks: Opportunities and Challenges, in *Proc. of 7th IEEE International Workshop*, Frankfurt, Germany, July 2002, pp. 283–290.

28. H. Harney and C. Muchenhirn, Group Key Management Protocol (GKMP) Architecture, RFC 2094, July 1997.

29. W. Hu, N. Bulusu, and S. Jha, An Anycast Service for Hybrid Sensor/Actuator Networks, The University of New South Wales (UNSW) CSE Technical Report 0331, October 2003.

30. D. Liu and P. Ning, Efficient Distribution of Key Chain Commitments for Broadcast Authentication in Distributed Sensor Networks, *Proceedings of the 10th Annual Network and Distributed System Security Symposium*, pp. 263–276, San Diego, CA, February 2003.

31. D. Liu and P. Ning, Multi-Level μ-TESLA, A Broadcast Authentication System for Distributed Sensor Networks, Submitted. Also available as Technical Report, TR-2003-08, North Carolina State University, Department of Computer Science, March 2003.

32. S. Basagni, K. Herrin, D. Bruschi, and E. Rosti, Secure Pebblenets, University of Texas, Dallas; downloadable from: <http://www.cs.huji.ac.il/labs/danss/sensor/adhoc/basagani$_2$001securepebblenets.pdf>

33. F. Hu, Y. Wang, and H. Wu, Mobile Telemedicine Sensor Networks with Low-Energy Data Query and Network Lifetime Considerations, *IEEE Transactions on Mobile Computing* (to appear in the issue of the Third Quarter in 2005).

34. F. Hu and S. Kumar, The Integration of Ad Hoc Sensor Networks and Cellular Networks for Multi-class Data Transmission, *Ad Hoc Networks Journal (Elsevier)*, accepted for publication in August 2004, now downloadable from Elsevier; see http://www.sciencedirect.com/science/journal/15708705

35. H. Harney and C. Muchenhirn, Group Key Management Protocol (GKMP) Architecture, RFC 2094, July 1997.

36. X.S. Li, Y.R. Yang, M.G. Gouda, and S.S. Lam, Batch Rekeying for Secure Group Communications, in *Proc. of 10th International World Wide Web Conference*, May 2001.

37. T. Park and K.G. Shin, LiSP: A Lightweight Security Protocol for Wireless Sensor Networks, *ACM Transactions on Embedded Computing Systems*, Vol. 3, No. 3, August 2004.

38. N. Li, J.C. Hou, and L. Sha, Design and Analysis of an MST-based Topology Control Algorithm, *IEEE Infocom*, 2003.

39. R. Barr, *JiST — Java in Simulation Time Users Guide*, Cornell University, March 19, 2004.

40. R. Barr, *SWANS — Scalable Wireless Ad Hoc Network Simulator Users Guide*, Cornell University, March 19, 2004.

41. Crossbow sensor network products: http://www.xbow.com.

42. OPNET simulation: http://www.opnet.com.

43. W. Siddui, SPECTRA: Secure Power-Efficient Cluster Topology Touting Algorithm, MS thesis, Computer Engineering Department, Rochester Institute of Technology, RIT Wallace Library, June 2005.

Chapter 9

Security Issues in Ad Hoc Networks

Ming Chuen (Derek) Wong, Yang Xiao, and Xu (Kevin) Su

Contents

9.1 Abstract

Mobile ad hoc network is an emerging area of mobile computing and is gradually becoming a new paradigm of wireless network. Its attractiveness includes easy deployment, fast network setting up, and less dependence on infrastructure. In addition, mobile ad hoc network is different from tradition networks in many ways, such as no fixed infrastructure for mobile switching, frequent changes in mobile node topology, and multiple hops between two nodes. The nature of ad hoc makes it difficult to deploy security mechanisms, and thus vulnerabilities of such a network naturally become exploitation of attackers with malicious intentions. In this chapter, we analyze demands, identify challenges, and survey security issues in mobile ad hoc networks. Secure requirements, attacks, secure routing, key management, and intrusion detection are discussed in detail.

9.2 Introduction

In mobile ad hoc networks, mobile nodes communicate directly with each other in a peer-to-peer manner and form a network on-the-fly, carrying out basic operations such as routing and packet forwarding without the help of an established infrastructure. Security is a serious concern to ad hoc networks due to them being much more vulnerable to malicious exploits than wired networks. Wired networks use dedicated nodes to support basic network operations such as routing, packet forwarding, and network management. In ad hoc networks, network operations are carried out by all available nodes, and such dependency is the fundamental reason for all the security problems that specifically belong to ad hoc networks.

Ad hoc networks have been used in many areas, ranging from personal area networks to military battlefield networks. There are many reasons that lead to the popularity of ad hoc networks [3]:

- *Cost savings*: Because there is no need to purchase or install access points of wireless LANs, a considerable amount of money is saved when deploying ad hoc networks.
- *Rapid setup time*: Ad hoc networks only require installation of radio NICs in users' devices. Therefore, the time to set up an ad hoc network is much less than that needed to install an infrastructure.

- *On-the-fly network:* Because there is no infrastructure needed in ad hoc networks, deployment is simple. A node can join a network anytime and build a network on-the-fly.

Nodes in an ad hoc network communicate in a single-hop or multi-hop fashion, where intermediate nodes between a pair of communicating nodes act as routers. Each node can operate as both a host and a router. Nodes can be highly mobile and, thus, creation of a routing path can be directly affected by mobility as well as additions and deletions of nodes. As a result, network topology can change rapidly, randomly, and unexpectedly.

During the past few years, lots of research has been done in ad hoc networks. Most of the research has focused on routing. Security issues have received relatively less attention, and these issues must be addressed.

There are many challenges related to deploying security mechanisms in ad hoc networks. Nodes have limited wireless transmission ranges so that communications between two nodes usually require a long routing path from the source node to the destination node. A longer routing path indicates a higher probability for packet losses due to transmission errors, frequent route changes, and potentially frequent network partitions.

Major security issues in ad hoc networks include vulnerabilities, security requirements, attacks, secure routing, key management, and intrusion detection. This chapter points out security problems and challenges of security mechanisms in security requirements and attacks, routing, key management, and intrusion detection systems. Routing is very different in ad hoc networks from that in wired networks because routing operations are carried by all available nodes, and there is no guarantee that the nodes do what they are supposed to do (e.g., forwarding packets). Secure routing can be categorized into two classes: *conservative approach* and *responsive approach*. In the conservative approach, malicious nodes are avoided intelligently (e.g., selecting a safer route). In the responsive approach, not only malicious nodes are avoided, but also actions are taken against malicious nodes (e.g., isolating misbehaved nodes from the network). To ensure confidentiality, some degree of cryptographic key is utilized for encrypting some important information, such as route messages. There are many challenges in establishing an efficient, robust, and distributed key management scheme, for suiting the nature of ad hoc networks. An intrusion detection system is used for detecting any anomaly among network nodes by analyzing the gathered network information. We introduce agent-based, distributed, and cooperative intrusion detection systems.

The remainder of this chapter is organized as follows. Vulnerabilities, security requirements, and attacks are presented in Section 9.3. In Section 9.4, secure routing schemes are discussed. Key management schemes are presented in Section 9.5. In Section 9.6, intrusion detection systems are introduced. Finally, we conclude this chapter in Section 9.7.

9.3 Vulnerabilities, Security Requirements, and Attacks

In the following subsections, we present vulnerabilities, security requirements, and attacks.

9.3.1 Vulnerabilities

An ad hoc network is much more vulnerable than a traditional wired network because of high mobility, no fixed infrastructure, and lack of central authority to enforce security mechanisms. There are three main vulnerabilities of an ad hoc network, as follows [1]:

1. Unlike a traditional wired network, no physical access is needed in an ad hoc network. Therefore, it is very easy for attackers to come to the network from all directions and to target any node in the network. Because there is no central infrastructure to enforce security mechanisms, every node must be prepared to encounter attacks directly or indirectly.

2. Nodes are autonomous units that are capable of roaming independently. Nodes with inadequate physical protection are receptive to being captured, compromised, and hijacked. Furthermore, tracking down a particular node in a large-scale ad hoc network may not be easily done, and attacks that come from within the network are usually more destructive and more difficult to detect. Therefore, the nodes and the network must be prepared to operate in a mode that trusts no peer.

3. Decision making in an ad hoc network is sometimes decentralized. However, some wireless network algorithms rely on the cooperative participation of all nodes. Lacking central authority makes an attacker easily exploit new attacks to break the cooperative algorithms.

9.3.2 Security Requirements

Security requirements are similar to those in other networks. The goal is to protect information and resources from attacks and misbehavior. To enforce security mechanisms, the following security requirements must be considered [2]:

- *Confidentiality* ensures that a transmitted message cannot be understood by anyone else and can only be accessed by the intended nodes.
- *Authentication* allows for communicating parties to know the identities of each other in order to make sure that they are genuine.
- *Integrity* ensures that the transmitted message is original and has not been altered throughout the transmission.
- *Non-repudiation* ensures that the transferred message has been sent and received by the parties claiming to have sent and received it.
- *Availability* ensures the availability of network services whenever it is required by the intended parties.

9.3.3 Attacks

Threats include attacks that intentionally, directly or indirectly, cause any damage to the network. Attacks can be classified as external attacks or internal attacks:

- *External attacks*: Malicious nodes involved in external attacks usually attempt to exploit authentication mechanisms to gain access to the network. After they gain access to the network, they will launch a second step attack — internal attack.
- *Internal attacks*: An internal attack is usually launched by an internal node in the network, and an internal attack is much more difficult to prevent, compared with an external attack, due to the nature of the ad hoc network.

Attacks can be also classified as passive attacks or active attacks:

- *Passive attacks*: Passive attacks usually involve eavesdropping packets during transmissions and analyzing them for further attacks. Because the wireless communication medium of an ad hoc network is widely shared, it is very easy to launch such attacks.
- *Active attacks*: Active attacks on the victims, such as resource deprivation, forces all routes to pass through a victim and leads to a shutdown of communication between nodes.

9.4 Secure Routing

Routing in ad hoc networks is very different from conventional wireless network in many ways. For example, there are neither centrally administered secure routers nor strict security policies. In addition, the nature of ad hoc networks is highly dynamic, and current ad hoc routing protocols are operating under the assumption that all participating nodes can be trusted, but unfortunately, this is not true. Routing can be defined as a mechanism of information exchange between two hosts in a network [6]. Routing is important because it eases communications between different parties in networks. Routing protocols in ad hoc networks can be categorized into three kinds: (1) table-driven/proactive protocols, (2) on-demand-driven/reactive protocols, and (3) hybrid protocols.

1. *Proactive protocols*: In proactive protocols, each node constantly updates routing information within a network, so that when a route is needed, the route information can be retrieved immediately. The merit is that a proactive protocol may have a lower latency because routes are maintained at all times, but the demerit is that a proactive protocol can possibly result in higher overhead due to constant route updates. The traditional link-state and distance-vector routing protocols in wired networks are proactive.

2. *Reactive protocols*: In reactive protocols, a route is created only when requested by a source node. When a source node requests a route to a desired destination, it initiates a route discovery process within the network. The process is completed only when at least one route is found or all possible routes have been examined. A route maintenance procedure maintains the discovered route and the route information until the destination becomes inaccessible or the route is no longer needed. The merit is that a reactive protocol may have lower overhead because routes are determined only if needed. The demerit is that a reactive protocol may have a higher latency because a route may be found only when it is being requested by a source node. Dynamic Source Routing (DSR) [7] and Ad Hoc On-Demand Distance Vector (AODV) [8] are two examples of reactive protocols.

3. *Hybrid protocols*: The Zone Routing Protocol (ZRP) is a hybrid protocol [16]. It attempts to incorporate the merits of on-demand and proactive routing protocols. The ZRP is similar to a cluster with the exception that every node acts as a cluster head but in the mean time, it also acts as a member of other clusters. The routing zone comprises a few nodes within one, two, or more hops from where the central node is formed. Each node has its own zone and does not rely on fixed nodes. Because it uses both reactive and proactive schemes, it exhibits better performance. However, because hierarchical routing is used, the path to a destination may be suboptimal. And because each node has higher level topological information, the memory requirement is larger.

9.4.1 Vulnerability of Ad Hoc Routing Protocols

Several protocols such as DSR or ADOV are based on flooding for data delivery from a source node to a destination node through intermediate nodes. There are some disadvantages to using flooding. First, a routing path may be too long, and too many nodes may receive data packets that need not be sent to them. Second, collisions may happen if multiple nodes send packets at the same time. Therefore, the longer the path, the less reliable it is because the chances of collisions and losing packets are higher.

According to Badache et al. [9], a reactive protocol is more suitable for ad hoc networks. Vulnerabilities of reactive routing protocols such as DSR and ADOV are listed as follows.

- *Naive trust*: Current routing protocols in ad hoc networks trust all nodes within the network and assume that all nodes will behave co-operatively and forward the packets requested. However, this naive assumption makes it easy for malicious nodes to insert erroneous routing updates [14,15], delay packet deliveries, replay old messages, change routing updates, or advertise incorrect routing information.

- *Misbehaving nodes*: Ad hoc networks maximize network bandwidth by using all available nodes for packet routing and forwarding. However, not all the nodes can be trusted, and misbehaving nodes can be a significant problem [5]. For example, a node can misbehave by agreeing to forward packets but failing to do it later.

- *Malicious modification*: In current routing protocols in an ad hoc network, all nodes are trusted by the network, and it is assumed that no nodes will modify protocol fields of messages passed among nodes. Routing protocols contain important messages such as route sequence number, hop count, or source routes. Because there is no central security mechanism to control nodes in a network, an enemy node or a compromised node can participate in routing packets and make malicious modifications to messages of routing protocols. Such an attack can cause redirection of network traffic and DOS attacks by simple modifications of routing protocol messages. For example, suppose that node F forwards a packet to node J, and then to destination node D. However, malicious node M can advertise itself to be a shorter route to node D by altering the message field in the routing protocol, and keep traffic from reaching node J. Such an attack is also known as a black hole attack [11].

- *Fabrication*: A malicious node can not only modify a message in the routing protocol, but also generate false messages [13]. For example, both ADOV and DSR have a path maintenance mechanism to recover broken paths when nodes are not available for routing packets anymore. Therefore, when the destination node or the intermediate nodes are not reachable, the upstream node broadcasts a route error message to all active upstream neighbors to notify them that a certain node is not available anymore. An attack can be launched by generating a false route error message to interrupt the communication between the nodes. For example, sender node S is trying to send a packet to destination node D, and malicious node J and malicious node K can launch a denial-of-service attack by continuously sending its neighbor route error messages. Node J sends it to node F and node K sends it to node G. Node F and node G will think that route error messages are generated by destination node D and they will delete the routing table entry for node D. Node F and node G forward a route error message to their upstream nodes, and these upstream nodes will also delete the route entry table for node D. As a result, node J and node K successfully prevent the communication between node S and node D.

- *Identity authentication:* As mentioned, current routing protocols in an ad hoc network trust every node in the network, and therefore there is not an authentication mechanism for Internet Protocol (IP) or Medium Access Control (MAC) address [12]. A malicious node can launch many attacks by changing its MAC or IP address. Because there is no identity authentication, the network cannot identify misbehaving nodes and remove or isolate them from the network. Both ADOV and DSR are vulnerable to this attack.

- *Information leakage:* A reactive protocol such as ADOV and DSR carries important information such as a route in cleartext format in a route discovery packet. An intruder can find out the structure of the network by analyzing the packet [13]. The information sometimes is very useful to intruders. For example, they can find out what are the adjacent nodes to the target or a physical location of a particular node. This information may be useful for many attacks.

9.4.2 Secure Routing Solutions

We survey several secure routing solutions for on-demand/reactive protocols such as DSR and ADOV as follows.

9.4.2.1 Security-Aware Ad Hoc Routing (SAR)

In Yi et al. [17], the Security Aware Ad Hoc Routing (SAR) protocol is introduced to find a secure route in the network, similar to the routing approach in Crawley et al. [18], but SAR uses different security attributes to improve the security of discovered routes, and integrates the security levels of nodes into the traditional routing metric mechanism. The objective of SAR is to represent a trust relationship between the nodes by measuring the trust level and by defining a suitable hierarchy of trust values. The representation of trust levels uses a simple hierarchy that reflects organizational privileges. The trust hierarchy can be numbers that reflect security levels of mobile nodes and paths associated with privilege levels. Integrating both the trust value of a node and security attributes of a route provides an integrated security metric for a discovered route.

In SAR, nodes are assigned trust values and embed into route request (RREQ) packets, and packets are routed only through the trusted nodes. For example, source node S requests packet delivery to destination node D. Before source node S broadcasts the packet, it specifies a desired level of security in RREQ. Assume that there are two shortest routing paths from source node S to destination node D: (1) node S ⇒ node E ⇒ node F ⇒ node J ⇒ node D, and (2) node S ⇒ node C ⇒ node G ⇒ node K ⇒ node D. If nodes C, E, and F fail to meet the security level specified by source node S, nodes C, E, and F must drop the RREQ and cannot forward the packet any further. Therefore, a route discovered by SAR may not be the shortest route, and SAR can only find a route with a quantifiable guarantee of security. If there is more than one route that satisfies the required security attributes, SAR finds the shortest one among them.

If there is no route that satisfies the requested security level desired by source node S, node S can modify the security attributes and send another RREQ to find a route with different security guarantees. If there is really no routing path that meets the security requirement, SAR fails to find a route even though the network may be connected.

Based on the authors of SAR, it is not sufficient to implement such a protocol because SAR heavily relies on the assigned trust value. However, the authors of SAR do not discuss how the trust value can be assigned or how the applications can determine the level of trust needed.

9.4.2.2 Secure Routing Protocol (SRP)

In Papadimitratos and Haas [19], the Secure Routing Protocol (SRP) is introduced. SRP provides correct routing information, where the correctness of the routing information requires a security association between the source and the destination node. SRP can also guarantee a node that a route discovery request will be able to identify and discard replies from misbehaving nodes or avoid receiving them. SRP uses a route field in RREQ and route reply (RREP) packets. Each intermediate node appends its identifier to the route field as a routing packet propagates from the source node to the destination node.

For example, assume that a route path from the source node S to the destination node D is S ⇒ node E ⇒ node F ⇒ node J ⇒ node D, as shown in Figure 9.1. Hence, there are four transmission hops involved in this packet delivery. In the first transmission, node S sends the packet to

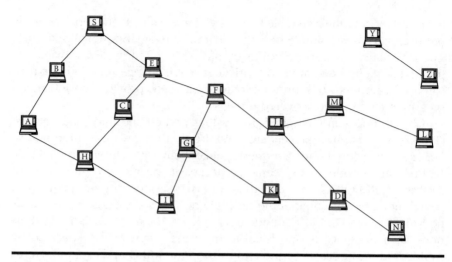

Figure 9.1   An ad hoc network: node S is the source node and node D is the destination node.

node E, and the route field will have the identity of node S; in the second transmission, node E sends the packet to node F, and the route field will have the identities of node S and node E; in the third transmission, node F sends to node J, and the route field will have the identities of node S, node E, and node F; in the fourth transmission, node J sends the packet to node D, and the route field will have the identities of node S, node E, node F, and node J.

However, SRP assumes that every node in the network is willing to provide its identity in the route field. Marshall [20] indicates that a false route can still occur by an invisible node attack; for example, a malicious node M located between node J and node L, shown in Figure 9.1. When node J forwards a packet to node M with the identity of J in the route field, node M can forward the packet to node L without putting the identity of node M. In other words, node M is basically invisible in the network and thus the route information from node M to node L is incorrect.

9.4.2.3   The Selfish Node (TSN)

In Buchegger and Boudec [21], The Selfish Node (TSN) is introduced. TSN attempts to identify and isolate misbehaving nodes, and the concept is taken from Dawkins theory [22]. In the biological world, birds groom parasites off each other's head because they cannot clean themselves. Dawkins divides birds into two types: "a sucker" that always helps, and "a cheater" that has other birds groom parasites off its head but fails to return the favor. Clearly, cheaters have advantages over suckers. Dawkins introduces a third kind of bird, a "grudger," that starts out to be helpful to every bird, but bears a grudge against cheaters and therefore no longer grooms their heads. Based on simulations provided by Dawkins, results show that cheaters will finally become extinct because the number of grudgers grows.

There are few components in TSN: a monitor (neighbor watch), a reputation system (node rating), a path manager, and a trust manager. Neighbors of one misbehaving node can detect the misbehaving node by listening to transmissions of the next node or by observing the route protocol behavior. The trust manager administrates a table of trust nodes and their trusted levels. The reputation system is a node quality rating system, which provides a means of obtaining a quality rating of participants in routing. The path manager performs path re-ranking according to a security metric, path deletion of path containing malicious nodes, action on receiving requests for a route from a malicious node, and action on receiving requests for routing a malicious node in the source route.

In TSN, if a node's noncooperative behavior has been detected and exceeds a threshold value, an ALARM message will be sent, and thus the misbehaving node will not receive packets making it isolated from the network.

For example, source node S is trying to send a packet to destination D. Assume that the path that it first chooses is S ⇒ node E ⇒ node F ⇒ node J ⇒ node D, as shown in Figure 9.1. However, node J does not forward the packet to destination node D. After the bad behavior of node J was observed by node F for a number exceeding a threshold, node F triggers an ALARM message to be sent to the source node S. Upon receipt of the ALARM message, node S acknowledges the message to the reporting node F and decides to use the alternate path.

However, TSN assumes that all nodes will be cooperative in observing the misbehavior nodes and send ALARM messages when observing misbehaviors of nodes. There is no guarantee that a misbehavior node will be reported.

9.4.3 Secure Position Aided Ad Hoc Routing (SPAAR)

In Carter and Yasinsac [23], Secure Position Aided Ad hoc Routing (SPAAR) is introduced. SPAAR uses locations to help in securing routing, and uses symmetric and asymmetric cryptography to encrypt RREQ. The current routing protocol is based on the flooding of the data packet, when a node initializes an RREQ, and many nodes receive packets that are not necessary. SPAAR uses position information to reduce the overhead of the route discovery process. However, the current position aided routing protocol is not secure because it passes location information in cleartext [23,24]. Therefore, SPAAR encrypts location information twice, first by destination public key and then by the group key among intermediate nodes.

There are two main components in SPAAR: (1) a neighbor table and (2) a route table. The neighbor table needs neighbor discovery and neighbor table maintenance, and the route table needs route discovery and route table maintenance. Each node is required to have a public and private key pair, called the neighbor group key pair. The private part is called the group encryption key (GEK) and the public part is called the group decryption key (GDK).

To add nodes to its neighbor table, node S periodically broadcasts a "Hello" message, and nodes within its range (node B and node E) shown in Figure 9.1 respond with a "Hello reply." Node S then distributes its GDK to each verified one-hop neighbor, where a one-hop neighbor is defined as if the distance is less than both nodes' transmission ranges.

In the route discovery in SPAAR, the source node first broadcasts an encrypted RREQ. Upon receiving an RREQ, an intermediate node decrypts the RREQ with the appropriate GDK and determines if it is closer than the node it received the RREQ from. If so, it forwards to its neighbors. This process is repeated until the RREQ reaches the destination.

Overall, SPAAR can ensure confidentiality within an ad hoc network. It uses the secure shortest geographical path. However, extra memory is required, because each node stores routing pairs, and there is heavy processing overhead for encryption. In addition, it needs extra hardware — GPS for locating node positions.

9.5 Key Management

In an ad hoc network, there is no authentication mechanism so that nodes can come and join the network and participate in packet routing without letting other nodes know who exactly they are.

Many threats target the above vulnerability, such as impersonation and denial of service. To address the identity authentication problem, encryption, authentication, and key management are widely used to prevent external attacks. Different types of keys can serve different kinds of purposes. For example, a public key is usually used for signing and encrypting messages to ensure the authenticity of individual nodes, as well as confidentiality and non-repudiation of the communications among mobile nodes. A shared secret key scheme can be used to encrypt messages to ensure confidentiality and to some degree the authenticity of routing information and data packets.

No matter what kind of security mechanism is used, some degree of cryptographic keys is needed to be shared among the communicating parties. As stated by Menezes et al. [27], the purpose of key management is to initialize system users within a domain; generate; distribute, and install keying material; control the use of keying material; update, revoke, and destroy keying material; and store, backup/recover, and archive keying material. There are some threats that must be dealt with by a key management system, which includes compromising the confidentiality of keys or the authenticity of the keys and unauthorized use of keys (e.g., the use of keys that are no longer valid).

9.5.1 Challenges

There are many challenges in key generation, distribution, and management. First, we must deal with dynamic topologies both in communication and in trust relationships. The assessment of whether to trust a wireless node can change over time. Second, we must deal with the lack of infrastructure support in an ad hoc network. It is widely recognized that an ad hoc network cannot rely upon centralized trust entities such as a key distribution center or certificate authority. In addition, there is a scalability problem; if each node in the network needs to share a secret key with every other node, then the number of keys that need to be managed is $n(n-1)/2$, which is very large if n is large.

9.5.2 Public Key Infrastructure (PKI)

In a public key infrastructure, each node carries a public key and a private key, where the public key is distributed to other nodes and the private key is kept secret and only known by itself. This type of key setup is essential for any service or application that employs asymmetric cryptography.

Figure 9.2 shows an example of components involved in deploying the public key infrastructure, where the end entity is the individual node of a certificate or the subject to which a certificate has been issued. Issuing and revoking the certification is being done by the certification authority (CA). A registration authority (RA) is an optional component because its function can be added to the CA. The RA is mainly responsible for establishing the identity of the owner of the certificate and associates the subject with its public key.

The basic operations of public key management scheme include [26]:

■ *Registration*: After a node creates its private key and public key, it needs to register its public key with the registration authority (RA), along with any information that is required for the certificate (e.g., name, email address, organization, etc.). Then the RA verifies the identity of the node. After the identification process is done, the RA contacts the certification authority (CA) and requests generation of the certificate.

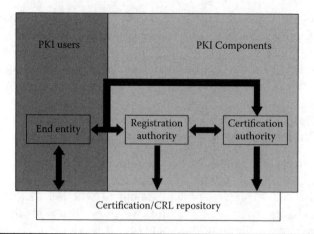

Figure 9.2   A public key infrastructure.

■ *Certification:* Certain information must be initialized before the node can use the services. One of the most important items is the public key that is registered and verified. After the initialization, the RA sends a certificate request to the CA, who generates and signs the certificate.

■ *Key maintenance:* Key maintenance includes key update and key revocation. In key update, the key pair becomes invalid after a period of time, and the key update service provides the transition to a new pair of keys and the issuing of the associated certificate. Key revocation is for maintaining the status of the certificates it has issued. Revocation is needed when the information of the certificate becomes invalid.

■ *Information distribution:* Information includes the certificate issued by the CA and revoked by the CA. After the CA issues a certificate, it needs to be made available to the owner and to others to use it. If a CA revokes a certificate, PKI must provide a mechanism that informs certificate users. A common method used is that the CA can publish a certificate revocation list (CRL) that lists all the certificates that have been revoked.

■ *Storage and authentication:* The nodes are responsible for storing the certificates. Each node stores its own certificate and all other nodes' certificates. The storage is fully distributed. When a node wants to obtain the authentic public key of another node, it first provides its own certificate repository, and then asks other nodes for their certificate repositories and finds an appropriate certificate chain.

9.5.3 Key Management Solutions

There are many proposed approaches for key management. Some authors use centralized authority or trusted third party for key management; some use public key infrastructure; some use a self-organizing approach. The following proposed solutions have been addressed explicitly in Fokine [29].

9.5.3.1 Partially Distributed Certificate Authority

Zhou and Haas [30] introduce a partially distributed certificate authority that uses a (k, n) threshold to set a certain number of server nodes, shown in Figure 9.3. The server nodes serve as certificate authorities, and each of the server nodes can only generate a partial certificate. A valid certificate can only be obtained by combining k such partial certificates.

Because the scheme is based on public key encryption, it requires that all the nodes be capable of performing the necessary computation. This solution is suitable for planned and long-term ad hoc networks because of their flexibility in adjusting the threshold based on the network size, and the robustness toward network scalability. However, the scheme assumes that s subset of the nodes is willing or able to take on the specialized server role.

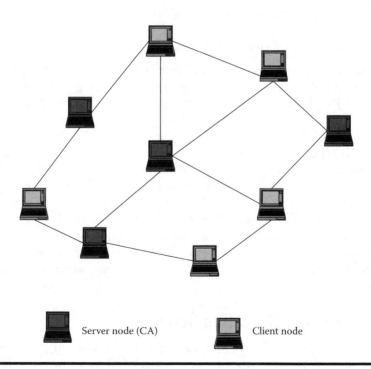

Server node (CA)   Client node

Figure 9.3   System architecture with four server nodes.

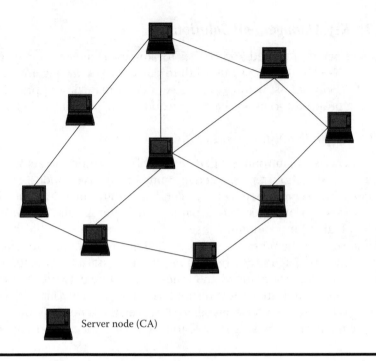

Server node (CA)

Figure 9.4   System architecture in which all nodes are CAs.

9.5.3.2   Fully Distributed Certificate Authority

Luo and Lu [31] extend the partially distributed certificate authority to a fully distributed certificate authority, as shown in Figure 9.4. It still uses the (k, n) threshold scheme to distribute an RSA certificate signing key to all nodes in the network. The major difference is that there is no need to elect or choose any specialized server nodes because the service is distributed among all the nodes. Therefore, once a node joins the network, it already has the capability to generate a partial key. The same with Zhou and Haas [30]; a valid key can only be obtained by combing k partial keys.

This solution can also be used in planned, long-term ad hoc networks with nodes capable of public key encryption. However, because a node has the capability of generating a certificate, it requires some computational capabilities from the participating nodes. In addition, the operation is under the assumption that every node will have a minimum of k one-hop neighbors.

9.6   Intrusion Detection Systems

It is important to have an intrusion detection system (IDS) for detecting inside and outside attacks. Inside attacks include unauthorized entrance/activity and file modification, whereas outside attacks can be denial-of-service attacks or attacks of penetrating a network infrastructure. Studies show that nearly all large corporations and most medium-sized organizations have installed some form of intrusion detection tool [34]. However, the current IDS are not strong enough to protect an ad hoc network.

The dynamic environment of an ad hoc network makes the design and implementation of an IDS difficult. There are few limitations of IDS in an ad hoc network [7]. First of all, there is no natural point for monitoring. Hosts may be disconnected at anytime, and the dynamically changing topology makes centralized analysis and correlation difficult. Furthermore, the IDS is just for detecting intrusion, and not for compensating for security weaknesses in network protocols.

According to Zhang and Lee [32], a good IDS should include the following requirements:

- *Run continually*: It must run continually without human supervision. The system must be reliable enough to allow it to run in the background of the system being observed. However, it should not be a "black box." That is, its internal workings should be examinable from outside.
- *Fault tolerant*: It must be fault tolerant in the sense that it must survive a system crash and not have its knowledge base rebuilt at restart.
- *Minimal overhead*: It must impose minimal overhead on the system. If the IDS will slow down the computer significantly, it should not be used.
- *Adaptability*: It must be able to adapt when the size of the network varies and the system behavior changes over time as new applications are being added. The IDS must be able to dynamically reconfigure itself to suit the dynamic environment.
- *Accuracy*: A good IDS should identify any anomaly when it is really happening and, oppositely, when the network has no intrusion, the IDS should not detect any anomaly. That is, the IDS should keep low false alarm and high detection rates.

There are two important assumptions of intrusion detection: (1) user and program activities are observable, through an examination of various parameters such as network traffic, CPU utilization, I/O utilization, user location, and various file activities [33]; and (2) attacks are different from "normal" (legitimate) activity and can therefore be detected by systems that identify these differences. Therefore, the basic operation of an IDS involves capturing the audit data, and then analyzing it and determine whether the system is being intruded. Therefore, the IDS should have the following three functions:

1. *Data auditing*: Data can be audited by continually monitoring activities (packet traffic or host behavior).
2. *Data analyzing*: Based on the audited data, automatically recognize suspicious, malicious, or inappropriate activities.
3. *Response*: When intrusion is being detected, trigger alarms to system administrator.

Based on the type of audit data, there are two basic types of intrusion detection [36]: host-based and network-based. Each has a distinct approach to monitor and secure data:

1. *Host-based IDS*: Examine data such as OS audits and system and applications logs held on individual computers that serve as hosts.
2. *Network-based IDS*: Examine data exchanged between computers such as the packet captured from the network traffic.

There are mainly two types of detection techniques equipped with IDS: (1) anomaly (behavior-based), which is any behavior outside a "normal profile" and (2) misuse (rule-based), in which monitored activity is compared to a set of signatures (patterns) for known attacks.

An agent-based IDS is described as follows. Due to the dynamic nature of an ad hoc network, intrusion detection and response in the network must be distributed and cooperative [32]. In this architecture, as shown in Figure 9.5, the left region of the network has a distributed IDS deployed, and every node equally participates in intrusion detection and response

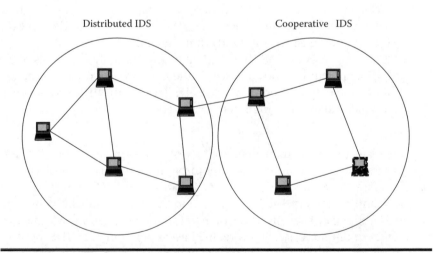

Figure 9.5 An IDS architecture with both distributed IDS (left) and cooperative IDS (right).

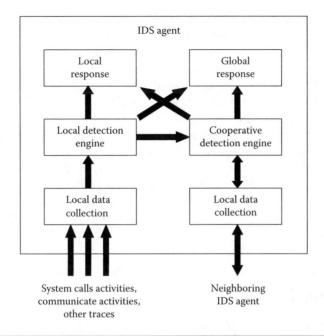

Figure 9.6   A conceptual model of an IDS agent.

for itself. The right region of the network deploys a cluster-based scheme, where a group of nodes in a cluster elect a node to be the monitoring node for the neighborhood. Each IDS agent runs independently and is responsible for detecting intrusions to the local node or its cluster. By combining these two schemes, the neighboring nodes can collaboratively investigate to not only reduce the chances of producing false alarms, but also detect intrusions that affect the whole or a part of the network. The ad hoc network is being defended by these individual IDS agents.

The internal part of an IDS agent, as shown in Figure 9.6, can be conceptually structured into six pieces [32]: the data collection module, the local detection engine, the cooperative detection engine, the local response and global response modules, and a secure communication module that provides a high-confidence communication among IDS agents:

- *Data collection*: The "local data collection" collects real-time audit data from various sources. It can be system and user application data, networking routing, and data traffic measurements.
- *Local detection*: The local detection engine analyzes the collected data and uses misuse or anomaly detection algorithms or other anomaly detection techniques to detect possible intrusions.
- *Cooperative detection*: If IDS detects an intrusion locally with a very high anomaly detection rate, then it can determine independently

that the network is under attack and can initiate a response. However, if a node detects an anomaly with a low anomaly detection rate, which needs broader investigation, it can initiate a cooperative global intrusion detection procedure. This procedure works by passing the intrusion detection state information to neighboring agents. If the agent(s) finds the intrusion evidence to be sufficiently strong, it initiates a response.

■ *Local and global response*: Intrusion response in an ad hoc network depends on the type of intrusion, possibly with the help from other security mechanisms if there is any, and the application-specific policy. For example, an intrusion response can be to re-authenticate the suspect nodes, re-initializing communication channels between nodes to or identifying the compromised node and reorganizing the network.

9.7 Conclusions

This chapter provided an overview of the existing security scenarios in mobile ad hoc networks, and showed that the nature of an ad hoc network has intrinsic vulnerabilities that cannot be removed. Secure ad hoc routing, key management, and intrusion detection aspects were discussed. Evidently, various attacks that exploit these vulnerabilities have been classified and studied. The current proposed solutions that attempt to solve security problems usually can only handle some very specific security problems, and there is no existing security mechanism that can handle a broad range of security vulnerabilities in ad hoc networks. However, new attacks constantly emerge, especially when ad hoc networks become more widely used. The ad hoc network security issue is important and still wide open for research with many open questions and opportunities for technical advances.

References

1. M. Prasant and K. Srikanth (Eds.), *Ad Hoc Networks Technologies and Protocols*, Springer, 2004.
2. W. Stallings, *Cryptography and Network Security: Principles and Practices*, 2nd ed., Prentice Hall, 1999.
3. J. Geier, *Wireless Networks First-Step*, Cisco Press, 2004.
4. C.E. Perkins, *Ad Hoc Networks*, Addison-Wesley, 2001.
5. I. Momhamod, *Wireless Networks*, FAU, CRC Press, 2003.
6. H. Bakht, Importance of secure routing in mobile ad-hoc networks, *Zata Computing — Unplugged Magazine*, August 2004.

7. B. David and A. David, Dynamic source routing in ad hoc wireless networks, *Mobile Computing*, Chapter 5, pp. 153–181, 1996.

8. C. Perkins and E. Royer, Ad hoc on-demand distance vector (ADOV) algorithm, *Proc. of the 2nd IEEE Workshop on Mobile Computing System and Applications (WMCSA'99)*, 1999, pp. 90–100.

9. N. Badache, D. Djenouri, and A. Derhab, Mobility impact on mobile ad hoc networks, *Proc. of ACS/IEEE Conference*, Tunis, Tunisia, July 2003.

10. S. Marti, T.J. Giuli, K. Lai, and M. Baker, Mitigating routing misbehaviour in mobile ad hoc networks, *Proc. of the Sixth Annual International Conference on Mobile Computing and Networking, MobiCom*, 2000.

11. F. Wang, B. Vetter, and S. Wu, Secure Routing Protocols: Theory and Practice, North Carolina State University, May 1997.

12. S. Kent and R. Atkinson, IP Authentication Header, RFC 2402, November 1998.

13. L. Zhou and Z.J. Haas, Securing ad hoc networks, *IEEE Networks Special Issue on Network Security*, November/December 1999.

14. A. Pirzada and C. McDonald, Establishing trust in pure ad-hoc networks, *Proc. of the 27th Australasian Computer Science Conference*, University of Otago, 2004.

15. S. Gupte and M. Singhal, Secure routing in mobile wireless and ad hoc networks," *Ad Hoc Networks*, 1 (2003) 151–174.

16. Z. Haas and M.R. Pearlman, "The Zone Routing Protocol (ZRP) for Ad Hoc Networks," draft-ietf-manet-zonezrp -02.txt, June 1999.

17. S. Yi, P. Naldurg, and R. Kravets, Security-aware ad hoc routing for wireless networks, *Proc. of the ACM Symposium on Mobile Ad Hoc Networking and Computing*, MobiHOC, 2001.

18. E. Crawley, R. Nair, B. Rajagopalanand, and H. Sandick, A Framework for QoS-based Routing in the Internet, RFC 2386, August 1998.

19. P. Papadimitratos and Z.J. Haas, Secure Routing for Mobile Ad Hoc Networks, *Proc. of SCS Communication Networks and Distributed Systems Modeling and Simulation Conference*, San Antonio, TX, 2002.

20. J. Marshall, An analysis of SRP for mobile ad hoc networks, *Proc. of the 2002 International Multi-Conference in Computer Science*, Las Vegas, 2002.

21. S. Buchegger and J. Boudec, The Selfish Node: Increasing Routing Security in Mobile Ad Hoc Networks, IBM Research Report RR 3354, May 2001.

22. R. Dawkins, *The Selfish Gene*, Oxford University Press, 1989 edition, 1976.

23. S. Carter and A. Yasinsac, Secure Position Aided Ad hoc Routing Protocol, *Proc. of the IASTED International Conference on Communications and Computer Networks (CCN02)*, Nov. 3–4, 2002.

24. S. Basagni, I. Chlamtac, V. Syrotiuk, and B. Woodward, A Distance Routing Effect Algorithm for Mobility (DREAM), *Proc. of the 4th International Conference on Mobile Computing and Networking*, Dallas, 1998, 76–84.

25. B. Karp and H. Kung, Greedy Perimeter Stateless Routing for Wireless Networks, *Proc. of the 6th International Conference on Mobile Computing and Networking*, Boston, 2000, 243–254.

26. C. Adams and S. Lloyd, Understanding Public-Key Infrastructure Concepts Standards and Deployment Considerations, August 1999.

27. A. Menezes, P. van Oorschot, and S. Vanstone, *Handbook of Applied Cryptography*, CRC Press, 1997.
28. R. Blom, An optimal class of symmetric key generation systems, *Proc. of EUROCRYPT 1984*, 335–338.
29. K. Fokine, Key management in Ad Hoc Networks, Master thesis. September, 2002.
30. L. Zhou and Z.J. Haas, Securing Ad Hoc Networks, *IEEE Networks*, Volume 13, Issue 6, 1999.
31. H. Luo and S. Lu, Ubiquitous and Robust Authentication Services for Ad Hoc Wireless Networks, Technical Report 200030, UCLA Computer Science Department, 2000
32. Y. Zhang and W. Lee, Intrusion Detection in Wireless Ad-Hoc Networks, *Proc. of the Sixth International Conference on Mobile Computing and Networking (MobiCom 2000)*, Boston, MA, August 2000.
33. T. Lunt, A survey of intrusion detection techniques, *Computers and Security*, 12, 4 (June 1993) 405–418.
34. SANS Institute staff, Intrusion Detection and Vulnerability Testing Tools: What Works?, *101 Security Solutions E-Alert Newsletters*, 2001.
35. P. Innella and O. McMillan, An Introduction to Intrusion Detection Systems, SecurityFocus.com, Infocus IDS article, 2006.
36. R. Bace, An Introduction to Intrusion Detection and Assessment: For System and Network Security Management, ICSA White Paper, 1998.
37. R. Power, 1999 CSI/FBI computer crime and security survey, *Computer Security Journal*, Volume XV, Number 2, 1999, p. 32.
38. P. Brutch and C. Ko, Challenges in Intrusion Detection for Wireless Ad-hoc Networks, SAINT Workshops, 2003, 368–373.
39. R. Bace, *Intrusion Detection*, Macmillan Technical Publishing, 2000.
40. S. Northcutt, *Network Intrusion Detection: An Analyst's Handbook*, New Riders, 1999.

Chapter 10

Security in Wireless Sensor Networks: A Survey

Wei Zhang, Sajal K. Das, and Yonghe Liu

Contents

10.1   Abstract

Wireless sensor networks have been rapidly emerging recently and the applications range widely. To protect the networks from different kinds of attacks, security in wireless sensor networks plays a crucial role and has received increased attention in the applications, especially those deployed in hostile environment such as battlefield monitoring and home security. However, the stringent resource constraints, such as energy, communication and computation capability, and memory, etc., pose many new challenges in security design. In this chapter we discuss the security issues in wireless sensor networks and present a comprehensive survey of the approaches published in recent literature. Some ongoing research work as well as open problems are also discussed.

10.2   Introduction

With recent advances in micro-electro-mechanical systems (MEMS) technology, wireless communications, and embedded systems, distributed wireless sensor networks (WSNs) are growing very fast. The applications range widely from military surveillance to civilian applications such as health monitoring [1]. A distributed sensor network is composed of a large number of self-organized sensor nodes and one or more base stations. The sensor nodes are in charge of sensoring, data processing, and routing. The base stations act as a gateway that connects the sensor network to the outside network. The sensor nodes communicate with each other using a wireless medium and collaborate together to accomplish some tasks. These battery-powered, low-cost sensor nodes are typically resource constrained, such as limited computation and storage capacity. The lifetime of the sensor nodes determines the lifetime of WSNs because it is infeasible to replace the batteries of these sensor nodes.

With the rapid emergence of applications, security in WSNs, especially in a hostile environment, is a key factor that can influence the entire network. In this chapter, we address the challenges of securing WSNs and overview the approaches in the literature that aim to protect WSNs from different attacks.

The remainder of this chapter is organized as follows. Section 10.3 discusses the security issues and assessment. Section 10.4 overviews the schemes for key management, followed by secure location information in Section 10.5. Section 10.6 presents the protocols for secure routing. Section 10.7 describes the mechanisms for broadcast authentication. Section 10.8 discusses secure aggregation. Some other security issues related to WSNs are discussed in Section 10.9, and we conclude the chapter and point out open problems in Section 10.10.

10.3   Security Goals and Challenges

10.3.1   Goals

Because sensor networks can be deployed in a variety of scenarios, security is an important issue in making the network work properly, especially in a hostile environment, such as battlefield monitoring and home security applications.

When talking about security, usually, the following are the goals that WSNs aim to achieve:

- *Confidentiality or privacy*: Ensure only authorized parties can access the data. That is, keep the information from disclosure to unauthorized parties.
- *Integrity*: Ensure only authorized parties can modify the data and the data is not altered during transmission.
- *Authentication*: Ensure the data is really sent by the claimed sender instead of fabricated by someone else.
- *Availability*: Ensure reliable delivery of data against denial of service.
- *Freshness*: Ensure the data is current and fresh (i.e., is not replayed by an adversary).

10.3.2   Attacks

The aforementioned security goals are not easy to achieve in WSNs. Some vulnerabilities inherent in sensor networks could be exploited to challenge these goals. All possible attacks against wireless networks could launch in WSNs. In particular:

- *Eavesdropping*: As in any other wireless communication, eavesdropping is easier in a wireless medium than in wireline networks. In WSNs, an adversary can access private information by monitoring transmissions between sensor nodes.

If eavesdropping is passive, only confidentiality is compromised. Using proper encryption of the messages can avoid this kind of attack. The situation exacerbates with active attacks, including participation and jamming. The following lists some attacks common in WSNs.

- *Compromising node*: In a hostile environment, the sensor nodes are subject to capture by adversaries. Once some nodes get compromised, all the information, even key information stored in it, might be exposed to the attacker. A desirable WSN should be able to detect the compromised nodes, exclude them, and update the asset affected by them.

- *Insertion node*: An attacker might add a node to the system and inject data as a legal node. Authentication could be used against this kind of attack, but not in all cases. A typical attack is the *Sybil* attack [63]. Under such attack, a single node can present multiple identities to control a substantial fraction of the whole network. It depreciates the effectiveness of the schemes that employ redundancy and also threatens geographic routing protocols in which the coordinate location information is very important to efficient routing.

In addition, by exploiting each layer's vulnerabilities of the network protocol stack, an adversary could launch a broad category of DOS attack that diminishes or eliminates the WSN's capacity to perform its expected function [40].

10.3.3   Challenges

A robust sensor network is desired to defend against all these attacks. However, the stringent resource constraints in sensor nodes and some characteristics of WSNs contribute to the challenges in security design [3].

- *Impracticality of public key cryptosystems*: The limitations of computation and power resource of sensor nodes impede the use of public key algorithms; thus, symmetric key cryptosystems are widely used in WSNs. A comprehensive survey investigating the computational requirements of a number of popular cryptographic algorithms (RC4, EC5, IDEA, SHA-2, and MD5) and embedded architectures (word size from 8 over 16 to 32-bit width) is presented in Ganesan et al. [24]. Although symmetric key cryptography is more suitable for sensor nodes, it complicates the key distribution and the broadcast authentication design.
- *Lack of deployment configuration knowledge*: Usually, the sensor nodes are deployed via random scattering by airplane so the sensor network protocols cannot know the topology of deployment. No prior knowledge of which nodes will be neighbors in the network calls into the key pre-distribution problem in WSNs.
- *Over-reliance on base stations*: Compared with sensor nodes, the base station has more resource in terms of power, computation, and storage capacity. Hence, the base station is treated as a trusted source and most security-related processing is done in the base station. However, attacks may launch on the base station and cause one-point failure.
- *In-network processing*: In WSNs, messages can be transmitted by multicast or flooding. The intermediate nodes need to access and

modify the message into a more compact message before further relay. This in-network processing/aggregation is an effective approach to reduce redundant messages and save energy. Because intermediate nodes need to access the contents of messages, pair-wise keys between sender and receiver cannot be used for encryption because this would preclude the intermediate nodes' access to the data. In-network processing requires a trust relationship beyond that which exists in traditional end-to-end security mechanisms. This property brings security concerns in relation to both integrity and confidentiality.

All these factors challenge security architecture design and require new solutions to handle them. For example, a study of the communication security aspects of WSNs is given in Slijepcevic et al. [26]. Based on the principle that data items must be protected to a degree consistent with their value, the authors assign different security levels to different types of data and propose a multitiered security architecture in which three types of data are classified based on SensorWare network architecture: mobile code, locations of sensor nodes, and application-specific data. Three security level mechanisms are defined corresponding to the types of data required to protect the data to a degree consistent with their value. This is a straightforward approach to secure communication in WSNs.

In general, due to the inherent properties of WSNs, a cross-layer design paradigm is employed to maximize the overall network performance by merging closely correlated protocols and the synergy between the various layers [50,51]. Figure 10.1 lists the layers of typical WSNs and the countermeasures to the possible attacks that might launch in the corresponding layer.

At the physical layer, the most common attack is jamming, which interferes in WSNs with the same RF that the network is using. The standard defense against jamming is spread-spectrum communication so that the jammers must either follow the precise hopping sequence or jam a wide section of the band to launch the attack [40].

At the MAC layer, before communication, to set up a secure and authenticated link between two sensor nodes, link layer shared-key-based cryptography is a typical defense against eavesdropping. In addition, after the deployment, the infrastructure must be established for the collaborative work. In particular, time synchronization and location information are two central factors in many applications. As a result, how to secure such information is vital to those applications.

At the networking layer, although routing is one of the most developed areas in WSNs, many proposed routing protocols have not been designed with security as a goal and they are vulnerable to different attacks [2]. Secure routing to ensure reliable message forwarding remains an open problem.

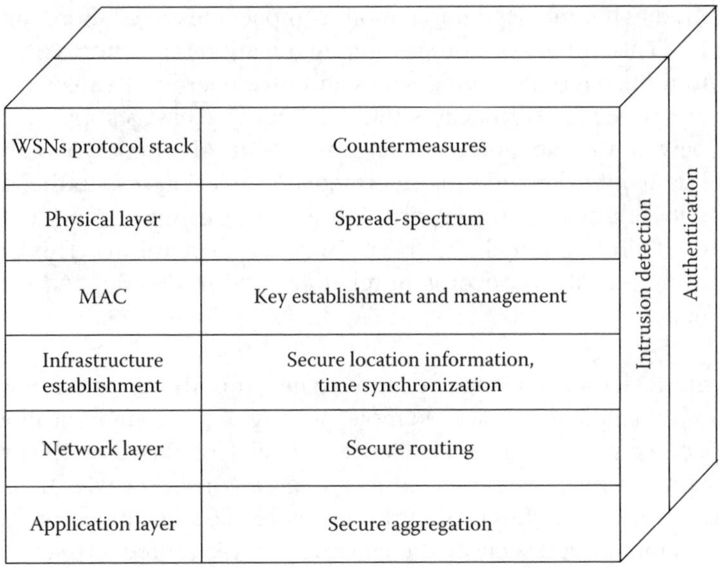

WSNs protocol stack	Countermeasures		
Physical layer	Spread-spectrum	Intrusion detection	Authentication
MAC	Key establishment and management		
Infrastructure establishment	Secure location information, time synchronization		
Network layer	Secure routing		
Application layer	Secure aggregation		

Figure 10.1   WSNs protocol stack and security defenses.

In general, WSNs are task specific and application dependent; however, no matter what kind of application is running, in-network processing is a common approach for energy savings but also brings security concerns especially under node-capture attack.

Additionally, WSNs are susceptible to different forms of intrusion, particularly node-capture attack. A compromised node is likely to reveal all its secret information to the attacker, including its secret keys. Therefore, an attacker can circumvent the purely cryptographic-authentication mechanisms and disrupt the WSN's normal operations. How to detect these compromised nodes is an intriguing challenge and needs collaborative work from different layers. As well, authentication is indispensable for all security designs.

Among the topics listed in Figure 10.1, some countermeasures, such as key establishment and management, have been extensively investigated while others have not. The following sections overview the related mechanisms from the bottom up.

10.4   Key Establishment and Management

Key establishment and management is an important security primitive and plays a pivotal role in other security services. However, how to set up secret keys between the sensor nodes in WSNs is nontrivial due to the limited resource in sensor nodes. Energy consumption in broad key management protocols and the analysis of the constraints, as well as the overhead of these protocols on a wide variety of hardware platforms, have been investigated in Carman et al. [25].

Traditionally, there are three types of general key agreement schemes: (1) trusted-server scheme, (2) self-enforcing scheme, and (3) key pre-distribution scheme [5]. The *trusted-server* scheme depends on a trusted online key distribution center (KDC) for key agreement between nodes. The potential problem in this scheme in WSNs is that sometimes the KDC may be not reachable for some sensor nodes due to communication range limitations. The *self-enforcing* scheme uses asymmetric cryptography, such as public key certificates, for key agreement. However, limited computation and power resources in sensor nodes often make public key algorithms infeasible. The third type of key agreement scheme is *key pre-distribution*, which means key information is distributed (stored in ROM) among all sensor nodes before deployment. Because no prior topology knowledge is available before deployment, the number of keys stored in each sensor node becomes a performance factor that affects the efficiency of key establishment in a pre-distribution scheme.

For the pre-distribution scheme, the simplest way is to use a single session key for all the sensor nodes so that all sensor nodes use the same key to encrypt/decrypt the messages. This has the least memory usage in each sensor node but it is obviously the least secure because the entire sensor network will be compromised once a single sensor node is compromised. The other way is to set up pair-wise private sharing keys between every two sensor nodes. This is much more secure because any capture of a node can only affect the link that the compromised node is directly involved in. However, this solution brings with it an implementation problem. Suppose there are N sensor nodes in the network; then, each sensor node has to store $N - 1$ keys and there are a total of $N(N - 1)/2$ keys in the network. This is impractical for the network composed of a large number of nodes. In addition, this scheme also gives rise to a scalability problem.

Because both of these extreme solutions are not desirable in realistic sensor networks, some trade-off approaches between the two have been developed. Based on the specific topology of the sensor network, key management can be further classified into pair-wise key and group key. The pair-wise key is only shared by every two different sensors, while the group key is shared by all sensor nodes in the same group.

10.4.1 Pair-wise Key

Inspired by the idea of probabilistic key sharing among nodes, Eschenauer and Gligor first present a random key pre-distribution scheme and use the shared-key discovery protocol for key distribution, revocation, and re-keying [4]. This scheme includes selective distribution, revocation of keys to sensor nodes, and re-keying.

For selective key distribution, before deployment, each sensor node is installed with a ring of keys that is randomly picked up from a large key pool (generated offline). Because of the random choice of keys on key rings, a shared key may or may not exist between every two nodes. Upon deployment and network initialization, shared-key discovery is performed. In this phase, the nodes find out if they share a key with their neighbors by either broadcasting the list of identifiers of keys or through a challenge-response protocol. A link exists between two sensor nodes if they share a key and all communication on that link is encrypted by this shared key. Otherwise, a path-key is assigned to selected pairs of nodes in a wireless communication range that do not share a key by using multi-link paths of shared keys. After key setup is complete, any pair of nodes has a certain probability of having at least one shared key, and thus a connected graph of secure links is formed.

Key revocation is performed whenever a sensor node is compromised. To effect revocation, a controller node (has a large communication range) broadcasts a single revocation message containing a signed list for the entire key ring to be revoked. After receiving the signature key, each node verifies the signature of the signed list of key identifiers, locates those identifiers in its key ring, and removes the corresponding keys. Re-keying takes place when some keys expire, which is equivalent to self-revocation of a key by a node.

One design issue concerns how to choose proper size for both the key ring and key pool so that the graph is connected with high probability (connected means that every pair of sensor nodes can establish a secret key directly or through a path). The length of the key-path is also an important performance metric for the efficiency of the design.

After the probabilistic key-sharing scheme for WSNs was introduced, based on it, some other works were proposed for improvement. We call this scheme the basic scheme in this chapter.

Chan et al. [3] generalize the basic scheme and propose three new mechanisms using the framework of random key pre-distribution. First, the *q-composite random key pre-distribution scheme*: to establish a secure link, instead of using a single sharing key from the key ring between two neighboring nodes, each node must discover q common keys it possesses with each of its neighbors. The key-setup will not be performed between nodes if they share fewer than q common keys. Second, the *multipath key reinforcement scheme* in conjunction with the basic scheme: suppose there are j disjoint paths between two sensor nodes A and B, and A first generates j random keys and then routes each random key along a different path to B. When B has received all j keys, a new link key will be computed with the link key k (single sharing key in basic scheme) and all j random keys.

The resilience of this scheme is strengthened because unless an adversary successfully eavesdrops on all j paths, he cannot know sufficient information to reconstruct the new link key. A link is considered completely compromised only if it is compromised and all its reinforcement paths are also compromised. Third, the *random pair-wise keys scheme*: a modification on the pair-wise key scheme to possess node-to-node authentication. Specifically, instead of storing all $N - 1$ keys, each node need only store a random set of Np pair-wise keys so that any two nodes can be connected with probability p. With this scheme, if any node is captured, the rest of the links not directly involved remain fully secure.

Each of these three schemes represents a different trade-off in the design space of random key protocols. Because an adversary must break all the q keys to get the message, the q-composite scheme achieves significantly improved security under small-scale attack while it is more vulnerable to large-scale attack because larger fractions of the network will be revealed to the adversary. The *2-hop* multipath reinforcement scheme improves security at the cost of increasing network communication overhead to find multiple disjointed paths. The random pair-wise scheme has the best resilience against node capture attacks and node-to-node authentication. However, it is not well-suited for large sensor networks because memory usage may be still too costly.

By combining the Blundo's polynomial-based key pre-distribution protocol [8] and key pool idea used in the basic scheme, Liu and Ning develop a general framework for pair-wise key establishment [7]. In Blundo et al. [8], to predistribute pair-wise keys, the key setup server randomly generates a bivariate $t - degree$ polynomial: $f(x, y) = \sum_{i, j=0}^{t} a_{ij} x^i y^j$ over a finite field with the property of $f(x, y) = f(y, x)$. Applying this in WSNs, for each sensor i, the setup server computes a polynomial share of $f(x, y)$, that is, $f(i, y)$. For any two sensor nodes i and j, node i can compute the common key $f(i, j)$ by evaluating $f(i, y)$ at point j, and vice versa. In this scheme, each sensor node stores a t-degree polynomial $f(i, x)$, and the key pool is composed of randomly generated bivariate polynomials. This scheme is a generalized basic scheme by replacing the simple key pool with bivariate polynomials key pools and has two special cases. When the polynomial pool has only one polynomial, the framework degenerates into the polynomial-based key pre-distribution. When all the polynomials are 0-degree, the polynomial pool degenerates into a key pool in the basic scheme.

Another variation of the basic scheme is proposed by applying Blom's key pre-distribution scheme on multiple key spaces [5]. Blom's scheme [6] is a deterministic scheme and allows any pair of nodes to find a secret pair of keys between them using $\lambda + 1$ (λ is much smaller than N) memory spaces in each node. This scheme has the λ-secure property: as long as no more than λ nodes are compromised, the uncompromised nodes are perfectly secure; when more than λ nodes are compromised, all pair-wise keys of the entire network are compromised. Blom's scheme is generalized in Du et al. [5] using multiple key spaces instead of just one key space for all nodes to compute the secret key. Specifically, first, ω spaces are constructed using Blom's scheme. Then, each sensor node carries key information from $\tau(2 \leq \tau < \omega)$ randomly selected key spaces. Two nodes can compute their pair-wise key if they share a common key space; otherwise, they can conduct key agreement via other nodes that share pair-wise keys with them. The essential idea of assigning some random keys to each sensor node is similar to the basic scheme, and it also follows the key pre-distribution and key discovery phase as in the basic scheme. Thanks to Blom's λ-secure property, this scheme holds a threshold property: when the number of compromised nodes is less than the threshold, the probability that any nodes other than these compromised nodes is affected is close to zero.

Combinatorial design theory is also novelly applied to the basic key pre-distribution scheme [13]. Some block design techniques (symmetric balanced incomplete block design and finite generalized quadrangle) of combinatorial design theory are mapped to key distribution to generate the key chains and the key-pools. The combinatorial scheme shows better connectivity with smaller key-chain size and decreases the key-path length as well.

By combining different techniques in cryptograph or mathematics, the above schemes aim to strengthen the security and improve the connectivity of the basic scheme. In addition to these two issues, communication efficiency is also a main concern in WSNs. In the key pre-distribution phase of the basic scheme, the key identifiers of a key ring and associated sensor identifier are stored in a trusted controller node. Each node must exchange the IDs of the keys to discover its neighbors that are communication-inefficient.

An improvement in communication overhead was proposed by trading computation for communication [11]. The design is based on the combination of two techniques: (1) probabilistic key sharing and (2) threshold secret sharing. Mainly, there are two differences with respect to the basic scheme. First, in the key pre-distribution phase, key assignment is ID-based so that no key identifiers exchange in communication. Specifically, the key setup server assigns each sensor node m distinct keys from the key pool of l keys. For each node with a unique node ID, using a pseudo-random number generator upon input of a node ID, the key server generates m distinct integers between 1 and l, and these m integers are the IDs of the keys for this node. So, a sensor node can deduce any other node key set ID as long as the nodes ID is provided. Furthermore, a node knowing the IDs of its neighbors can determine not only which neighbors share or do not share keys with it, but also which two neighbors share which keys. This knowledge leverages logical path discovery. Another difference is that the session key between any pair of nodes is exclusively known to these two nodes. In the basic scheme, because the key ring for each sensor node is randomly selected from the key pool, it is possible that a session key is known by several sensor nodes instead of only the two involved ones, and this compromises authentication. To avoid this, a sender node splits the to-be-established pair-wise secret key into multiple shares using an appropriate secret-sharing scheme. The sender then transmits to the recipient node all these shares over multiple different logical paths for each share. The recipient node then reconstructs the key after it receives all (or a certain number of) the shares and sends back a HELLO message. The key establishment process is finished after the sender verifies the HELLO message.

Pietro et al. [12] also address the communication overhead in the shared-key discovery phase of the basic scheme, and adopt the pseudo-random seed-based key deployment strategy introduced in Zhu et al. [11] to present two protocols for key establishment: Direct Protocol and Cooperative Protocol. In the Direct Protocol, suppose sensor node A wants to communicate with sensor node B. First, A computes all the keys it shares with B using a pseudo-random number generator on the indexes of the keys of B. Then, the keys are *XOR-ed* together. The resulting value will be used to secure all messages between A and B. Sensor B, once it receives the first message from A, builds $k_{a,b}$ in the same manner to decrypt the message. In the construction of key $k_{a,b}$, the protocol employs all the shared keys between sensors A and B. In the Cooperative Protocol, if sensor A wants to establish a secure channel with sensor B, sensor A chooses a set $C = \{c_1, \ldots, c_m\}$ of cooperating sensors such that $A, B \notin C$ and $m = 0$. Then, A sends a request of co-operation to each of the sensors in C, and the session key between A and B is obtained with information of each $c \in C$. A nice property of the Cooperative Protocol is that because sensor A can add/remove some sensors to C to achieve a satisfactory level of security, the network security properties can be dynamically changed during the lifetime. Figure 10.2 shows how the Cooperative Protocol works.

The basic scheme and all the above improvements are under a common assumption that no deployment knowledge is available and the sensor nodes are distributed uniformly over the entire deployment area. Du et al. [9] derive the spatial relation between sensors prior to deployment and present a key pre-distribution scheme based on the deployment knowledge model. Based on the fact that sensor nodes are usually deployed in groups, a group-based deployment model is developed. In this model, N sensor nodes to be deployed are divided into $t * n$ groups of equal size, and each group is arranged in a grid for deployment. The actual final location of a sensor node in a particular group follows some nonuniform distribution (e.g., two-dimensional Gaussian distribution in [9]). The location relation between sensor nodes (which nodes are more likely to be close

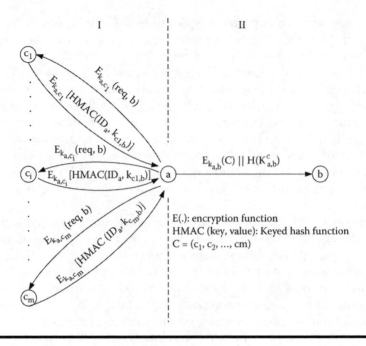

I   II

$E_{k_{a,c_i}}(req, b)$

$E_{k_{a,c_i}}[HMAC(ID_a, k_{c1,b})]$

$E_{k_{a,c_i}}(req, b)$

$E_{k_{a,c_i}}[HMAC(ID_a, k_{c1,b})]$

$E_{k_{a,b}}(C) \| H(K^c_{a,b})$

$E_{k_{a,c_m}}(req, b)$

$E_{k_{a,c_m}}[HMAC(ID_a, k_{cm,b})]$

E(.): encryption function
HMAC (key, value): Keyed hash function
C = (c₁, c₂, ..., cm)

Figure 10.2   Illustration of the Cooperative Protocol.

to each other) can be derived from the deployment distribution. A key pre-distribution scheme, which is based on the basic scheme and takes advantage of deployment knowledge, is proposed. This scheme is composed of three phases: (1) key pre-distribution, (2) shared-key discovery, and (3) path-key establishment. The latter two phases are exactly the same as in the basic scheme, and the deployment knowledge contributes the difference in the first phase. First, the key pool S is divided into $t*n$ key pools $S_{i,j}$, each of which corresponds to a deployment group. Then, each sensor node in the deployment group is randomly assigned some number of keys from its corresponding key pool. The deployment knowledge renders the goal that the nearby key pools (their corresponding deployment groups are neighbors or nearby) share more keys, while pools far away from each other share fewer keys or no keys at all. Specifically, by defining overlapping factors a and b, the scheme has (1) two horizontally or vertically neighboring key pools sharing exactly $a|Sc|$ keys; (2) two diagonally neighboring key pools sharing exactly $b|Sc|$ keys; and (3) two non-neighboring key pools sharing no keys. This is illustrated in Figure 10.3.

Independent of previous work, another key pre-distribution scheme with location knowledge was proposed by Liu and Ning [10]. It is noticed that in static sensor networks, although it is difficult to precisely pinpoint sensors positions, it is possible to approximately determine their

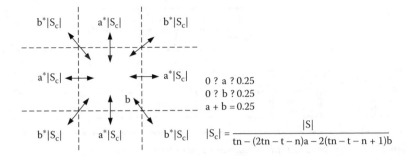

Figure 10.3   Pre-distribution based on deployment knowledge.

locations. Based on this observation, the authors present two pair-wise key pre-distribution schemes: (1) a closest pair-wise keys pre-distribution scheme and (2) a location-based keys scheme using bivariate polynomials. In both schemes, it is assumed that the key setup server is aware of the network-wide signal range and the expected location of each sensor before deployment as well as the deployment error. For the *closest pair-wise key scheme*, the basic idea is to have each sensor share pair-wise keys with c other sensors whose expected locations are closest to this sensor (c is a system parameter). For example, for each sensor u, the setup server first discovers a set of c other sensors whose expected locations are closest to the expected location of u. For each sensor within this set, the setup server randomly generates a unique pair-wise key if no pair-wise key between these two nodes has been assigned. Similarly, the other scheme, *location-based keys scheme* using bivariate polynomials, is an improvement upon their previous work [7] by taking advantage of location knowledge. The essential idea is to combine the closest pair-wise keys scheme with the polynomial-based key pre-distribution technique. Specifically, the deployment area is partitioned into small areas called cells, each of which is associated with a unique random bivariate polynomial. Then, instead of assigning each sensor the pair-wise keys for the closest sensors, each sensor is assigned a set of polynomial shares that belong to the cells closest to the one that this sensor is expected to locate in.

These schemes show that with the help of deployment or location knowledge, each node only needs to store a fraction of the keys required by the basic scheme while achieving the same level of connectivity. This improvement is twofold: it relieves the memory requirement and improves network resilience against node capture.

Figure 10.4 summarizes the relationship between the basic scheme and all the follow-ups; the numbers in square brackets indicate the references.

Furthermore, if the routing information is aware, it will greatly facilitate the key distribution (key update) in multicast communications. By

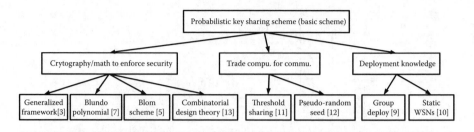

Figure 10.4   Pair-wise key pre-distribution schemes: a taxonomy.

combining routing and physical layer algorithms, Lazos and Poovendran [27] present a greedy, routing-aware key-distribution algorithm to secure broadcast communications.

Different from the above works, which only use off-line key setup servers before sensor node deployment, Park and Shin [39] propose a protocol called "LiSP" (Lightweight Security Protocol) for WSNs using online key servers. LiSP is composed of intrusion detection, group-key management, and message-key management. The message-key management periodically renews the message key and is the core of the protocol with some salient features, such as efficient message-key broadcasting without retransmissions/ACKs, implicit authentication for message keys without increasing additional overhead, fault tolerance in terms of detecting/recovering lost message keys, seamless message-key refreshment without disrupting on-going data transmission, and robustness to clock skews among nodes. In fact, LiSP is more a protocol suite than a simple key establishment scheme.

All the key pre-distribution schemes discussed above need either off-line trusted third parties to pre-load the keys in the initialization phase or on-line key setup servers. Chan [19] proposes the first Distributed Key Pre-distribution Scheme (DKPS) without relying on any trusted third parties and an established routing infrastructure. The basic idea of DKPS is that each node individually packs a set of keys from a large publicly known key space, by following some procedure; at the end, all nodes hold the exclusion property with a high probability that any subset of nodes can find from their key patterns at least one common key not covered by a collusion of, at most, a certain number of other nodes outside the subset. There are three phases in the DKPS scheme, namely (1) distributed key selection (DKS), (2) secure shared-key discovery (SSD), and (3) key exclusion property testing (KEPT). The scheme works as follows. Each node randomly picks keys into its key ring in the DKS phase, then uses the SSD protocol to find out which keys it has are in common with another node, and finally tests whether its keys satisfy the exclusion property. The SSD phase, based on Rivest's Privacy Homomorphism (PH) encryption functions, is the subtlest part of the scheme. In the key discovery phase, each node needs to

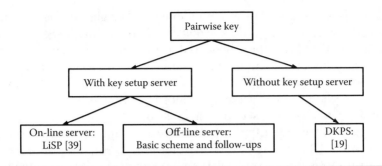

Figure 10.5   Pair-wise key schemes: a taxonomy.

find out with each one of the others which keys in their key rings are in common but not to leak any information about the keys that the other side does not have. Another nice property is that unlike most key agreement schemes, rekeying is not needed in the scheme when a member leaves or a set partition occurs. And it can work well before any routing infrastructure has been established. The major strength of DKPS is that no trusted third party is needed and it works without relying on any routing infrastructure. However, because cover-free family (CFF) construction is used to achieve perfect secrecy, the key storage requirement at each node may be large.

The summarized classification of pair-wise key schemes is shown in Figure 10.5.

10.4.2   Group Key

Because WSNs are often organized as hierarchical cluster architecture, group key is more efficient to secure the communications within one cluster. However, regardless of the centralized or distributed solutions, those group key management schemes proposed in wireline networks are not applicable in WSNs due to either the resource constraints in sensor nodes or the scalability of the networks.

To overcome the drawbacks, a family of distributed and localized group re-keying schemes, called the *pre-distribution and local collaboration-based group re-keying (PCGR)* schemes is proposed in Zhang and Cao [45]. In this scheme, the future group keys are preloaded to the sensor nodes before deployment in the form of *group key polynomials (g-polynomials)*. After deployment, each sensor node randomly picks an *encryption polynomial (e-polynomial)* to encrypt its *g-polynomial* and distributes some share of this *e-polynomial* to its neighbors, then removes both the *g-polynomial* and *e-polynomial*. In this way, a node cannot access its future group keys without collaborating with a certain number of neighbors. Later, the group key updating mechanism guarantees that a node can compute its new group key as long as it is trusted by a ceratin number of neighbors; meanwhile, the node cannot derive any group key that should not be disclosed at this time.

The essential idea of the scheme is that a sensor node cannot retrieve the group key without gaining a certain number of its group members' trust. Because the compromised nodes are likely to be first detected by their neighboring nodes, this scheme provides the promptest reaction possible to the ongoing changing security circumstance.

Another independent work to achieve the same goal is proposed in Chadha et al. [46]. In addition to the pre-distributed personal secrets and broadcast information, local collaboration among sensor node themselves is required in the scheme. To be more specific, a sensor node is not able to obtain the secret key solely based on the broadcast message and its pre-deployed secret share. Rather, it must seek for collaboration from its fellow sensor nodes. Only by jointly exploiting the secret shares disclosed by the broadcast message, its own pre-distributed secret, as well as secrets revealed by other nodes, can a node reconstruct the key. Different from the previous work, where the secret share (*e-polynomial*) possessed by a node is disseminated to the neighbors during the bootstrapping phase, in this work, the construction of the secret polynomials is done via the sink and hence true broadcast-based key distribution can be achieved. In addition, a set of enhancements is presented, including a self-healing scheme to accommodate the lossy nature of the wireless medium and a self-evolving scheme to allow sensor nodes to advance their personal secrets from session to session to reduce the memory requirement.

10.5   Secure Location Information

Location information in a WSN is sensitive information and the prerequisite for geographic routing; thus, it is also subject to attack.

There are two directions in secure location information. One focuses on how to compute the sensor node's position correctly even under different attacks. The other focuses on how to verify the sensor node's location claims.

10.5.1   Secure Location/Position Computation

In WSNs, GPS is the most common approach for obtaining geographic position information in outdoor applications. However, for civilian GPS, no secure protection is provided; thus, it is vulnerable to spoof attacks from GPS satellite simulators. Kuhn [54] proposes an information-hiding-based asymmetric security mechanism against the selective-delay attack and signal-synthesis attack on trusted positioning receivers.

A cryptographic-based scheme, SeRLoc (Secure Range-independent Location), is proposed in Lazos and Poovendran [53] to enable the sensors to determine their location even in the presence of malicious adversaries. In SeRLoc, based on directional antennas, each locator that acquires the location and the orientation through GPS receivers transmits different beacons at each antenna sector containing its coordinates and the angles of the antenna boundary lines with respect to a common global axis. Based on the beacon information, the sensors can determine their location by collecting the beacons from all locators they can hear and executing some algorithm. To protect the localization information, a security mechanism is integrated into the scheme. A global symmetric key is shared between sensors and locators. In addition, every sensor shares a symmetric pair-wise key with every locator so that the beacons from each locator can be authenticated to prevent impersonation via a one-way hash function. The analysis shows that SeRLoc is robust to several attacks, including the wormhole attack, Sybil attack, and compromised sensors. However, it does not consider the attacks of frequency jamming, and it assumes that locators are always trusted and cannot be compromised by an adversary.

Observing the fact that a cryptographic-based scheme cannot work when some sensor nodes get compromised, Liu et al. [56] propose an attack-resistant location estimation scheme that can survive malicious attacks even if the attacks bypass authentication. Two approaches are presented to deal with the malicious attacks. The first is based on minimum mean square estimation (MMSE) and uses the mean square error as an indicator to identify and remove malicious location references. The algorithm first estimates the sensor's location with the MMSE-based method and assesses if the estimated location could be derived from a set of consistent location references. According to the threshold-based consistency criteria, the estimation result is accepted or the above processing is repeated after the most "inconsistent" location reference is removed. The second method, voting-based location estimation, quantizes the deployment field into a grid of cells and has each location reference to "vote" on the cells in which the node can reside. Both location estimation techniques can tolerate attacks against range-based location discovery even if some compromised sensor nodes exist.

Location computation also can be secured by employing some techniques from other fields. In Li et al. [52], robust statistical methods are developed to make localization attack-tolerant. Two of the most common used schemes to localize a single device — triangulation and RF fingerprinting — are examined. For triangulation-based localization, instead of using the least square method, the least median of square position estimator is proposed. Similarly, robustness to fingerprinting localization is realized through the use of a median-based distance metric. By exploiting the redundancy represented in typical WSNs deployment, both these schemes are robust to outliers, which is dramatically different from the true value generated either by malicious attacks or non-adversarial corruption of measurement data.

10.5.2 Location Verification

Instead of directly secure the location computation by a node, another approach is to verify a device's position; that is, the node is in a region of it claims.

The first distance bounding technique that could be integrated with public key identification schemes is proposed by Brands and Chaum [57]. Using a time-bounded challenge-response protocol, the verifying party can determine a practical upper bound on the physical distance to a proving party and defend against man-in-the-middle attacks (*mafia frauds*).

A simple protocol without the requirement of cryptography and tight time synchronization, called "Echo," is proposed in Sastry et al. [38] to secure location verification in sensor networks. By exploiting the physical properties of sound and RF signal propagation in wireless communication, a set of verifiers can verify whether a prover is in a region of interest. The advantage lies in that the verification protocol can work with any location determination algorithm, even a potentially insecure one, without compromising the security of the ultimate guarantee that a prover is in the region.

A mechanism for both secure position computation and verification of positions, called Verifiable Multilateration (VM), is proposed in Capkum and Hubaux [58]. According to the distance bound measurements of time-of-flight from multiple reference points, an authority (base station) performs subsequent computations to get the correct claimant's position. Relying on the classical three-pass authentication protocols, the base station can also verify a position reported by the node. As an application of VM, SPINE (Secure Positioning In sensor NEtworks) system is presented to ensure secure positioning of sensors even in the presence of adversaries.

Another work that aims to provide robust position determination and location verification can be found in Lazos et al. [61].

In addition to these works, how to detect the malicious beacon nodes in WSNs is investigated in Liu et al. [59]. A suite of techniques is introduced to detect and remove compromised beacon nodes that supply misleading location information to the regular sensors. By checking the consistency between the distance estimation derived from measurement of the received signal strength indicator and the computation based on the known location and the claimant's position, malicious beacon signals can be detected. Furthermore, with the help of the wormhole detector, the source of this signal (either by a malicious node or a replayed one through a wormhole) can be testified. This work can be considered an intrusion detection mechanism in secure location information.

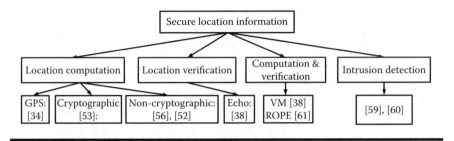

Figure 10.6   Secure location information: a taxonomy.

Another intrusion detection work with the same goal is presented in Du et al. [60] and takes advantage of deployment knowledge.

A classification for secure location information is shown in Figure 10.6.

10.6   Secure Routing

The in-network processing characteristic of sensor networks requiring intermediate nodes to have access to the data complicates the design of routing protocols. Once one of these intermediate nodes is compromised, it can eavesdrop and even modify the data, thus threatening the entire network. So, the routing protocols in sensor networks should provide not only reliable delivery, but also security services, especially for integrity and authentication. At the same time, a secure routing protocol should be robust against different attacks, such as denial of service, compromised nodes, etc.

The routing security problem in WSNs is first discussed in Karlof and Vagner [2]. This work summarizes attacks against the current proposed routing protocols and discusses countermeasures and design considerations for secure routing protocols. The attacks can be classified into two categories: (1) trying to manipulate user data directly or (2) trying to affect the underlying routing topology. Both kinds of attacks can consume valuable resources to cause a denial-of-service attack. It is claimed that most outsider attacks (attackers have no special access to the sensor networks) against sensor network protocols can be prevented by encryption and authentication using a cryptographic key. However, this is not effective in the presence of insider attacks or compromised nodes. And it is unlikely that there exist effective countermeasures against those attacks after the design of a protocol has completed. So, it is crucial to consider security issues at the beginning of routing protocol design.

The wormhole is one of the most dangerous attacks in WSNs. In a wormhole attack, a packet received by an adversary is tunnelled to another point in the network and replayed from that point. Depending on the adversary's behavior, the packet could be discarded or selectively forwarded or modified. In Hu et al. [55], a mechanism called packet leashes is proposed to detect and defend against the wormhole attack. A leash is any information that is added to a packet designed to restrict the packet's maximum allowed transmission distance. Specifically, two leashes are defined. A *geographical leash* ensures that the recipient of the packet is within a certain distance of the sender, and a *temporal leash* ensures that the packet has an upper bound on its lifetime, which restricts the maximum travel distance. By checking if the packet travels further than the leash allows, either type of leash can prevent the wormhole attack.

INSENS (INtrusion-tolerant routing protocol for wireless SEnsor Networks) aims to improve the routing robustness so that a single compromised node can only disrupt a localized portion of the network and cannot cause widespread damage in the entire sensor network [20]. It can provide protection against two classes of attacks: (1) DOS attacks and (2) routing attacks that propagate false routing information throughout the network. The routing is based on an asymmetric architecture composed of a base station and sensors. Each node shares a secret key only with the base station, and not with any other nodes. To prevent DOS attacks, individual nodes are not allowed to broadcast to the entire network. Broadcast by a base station is authenticated to prevent sensor nodes from spoofing the base station. To prevent false routing data, controlling information must be authenticated. To combat the compromised nodes, redundant, disjoint multipath routing is built into INSENS so that even if an intruder breaks down a single node or path, secondary paths will exist to forward the packet to the correct destination. Specifically, to facilitate communication between sensor nodes and a base station, INSENS constructs forwarding tables at each node in three rounds. In the fist round, the base station floods a request message to all the reachable sensor nodes, and a one-way sequence and keyed MAC algorithm is used to defend against intrusions. In the second round, sensor nodes send their local topology information using a feedback message to the base station in which, after verification, the messages that reach the base station are guaranteed to be correct and secure from tampering. In the third round, the base station computes the forwarding tables for each sensor node based on the information received in the second round and sends them to the respective nodes using a routing update message. In this way, the base station can collect all the connectivity information and authenticate it. All these heavy-duty computations are performed at the base stations to reduce computation at the senor nodes.

In addition to the sensor network's specific routing schemes, some secure routing mechanisms designed for ad hoc networks can also be adopted in WSNs.

SEAD (Secure Efficient Ad hoc Distance vector routing protocol) is a secure routing protocol on top of Destination-Sequenced Distance-Vector (DSDV) ad hoc routing protocol [21]. One-way hash functions and symmetric cryptographic operations are used in this protocol. To avoid attacks that might not use the average weighted settling time for delay in sending triggered updates, SEAD does not use such a delay as in DSDV. Unlike DSDV, when a node detects that its next-hop link to some destination is broken, the node does not increment the sequence number for that destination in its routing table to bypass authentication. Moreover, the lower bound on each metric in a routing update in SEAD is secured through authentication by one-way hash chains. In addition, the receiver of SEAD routing information authenticates the sender to ensure that the routing information originates from the correct sender. The source of each routing update message in SEAD is also authenticated by either TESLA or MAC (Message Authentication Code). SEAD is robust against multiple uncoordinated attackers creating incorrect routing states in any other nodes, despite active attacker or compromised nodes in the network.

Unlike SEAD, a periodic protocol in which nodes periodically exchange routing information with other nodes, Ariadne is an on-demand secure routing protocol for ad hoc networks: a node attempts to discover a route to some destination only when it has a packet to send [22]. The similarity between these two schemes lies in that both protocols are robust to some nodes being compromised and employ symmetric cryptography. A difference is that Ariadne is based on Dynamic Source Routing (DSR) protocols and combines TESLA to authenticate routing messages. The goal of Ariadne is to ensure that each node that interprets routing information must verify the origin and integrity of that data. For routing discovery, two stages are performed: (1) the initiator floods the network with a ROUTE REQUEST, and (2) the target returns a ROUTE REPLY. To secure routing against attacks, Ariadne provides the following properties: (1) the target node can authenticate the initiator's ROUTE REQUEST message by a MAC with a key shared between the initiator and the target over a timestamp; (2) three alternative mechanisms (TESLA, digital signature, or MAC) can be used for the initiator to authenticate ROUTE REPLY messages; and (3) a per-hop hashing technique to verify that no node can be removed in the node list in the REQUEST. For route maintenance, if a previous route is not available anymore, a ROUTE ERROR message will be returned to the original sender, and this message is also required to be authenticated using TESLA. Ariadne can provide protection against routing misbehavior as well as malicious route request floods. However, although it is claimed that Adriadne is designed to support nodes with few resources, those nodes such as Palm Pilot or RIM pager are still more powerful compared with sensor nodes.

In addition to securing the sensor nodes in WSNs, some efforts focus on improving the intrusion tolerance against attacks to isolate or destroy the base station [23,34]. Assuming that the network framework is composed of sensor nodes that are organized in a tree-like network routing structure around each base station and every sensor node processes the sensed data from all of its child nodes and itself, and sends the result to its parent node, two strategies are proposed against these attacks on base stations. First, secure multi-path to multiple destination base stations are designed to provide intrusion tolerance against isolation of a base station; a mechanism enables the secure setup of multiple routing paths to multiple base station, so that even an adversary can attack and destroy part of the sensor network, and the rest of the network can survive and continue to report data. Second, anti-traffic analysis strategies are proposed to help disguise the location of the base station from the eavesdropper. Most of the work is similar to that in Deng et al. [20].

Another branch of secure routing derived from ad hoc networks employs reputation obtained by observing the nodes' behavior as the criteria to guide routing. In general, a node can monitor its neighbors' activities to learn if they drop packets constantly and assign them different reputations based on their behaviors. Those misbehaving nodes could be identified once their reputations fall below some threshold. The routing path is established among the nodes that have a high reputation. For example, in Tanachaiwiwat et al. [62], trust routing for location-aware sensor networks (TRANS) is proposed for establishing trust routing by excluding the misbehaving nodes based on the blacklists. Two components — trust routing and insecure location avoidance — comprise the mechanism on top of geographic routing. According to the replies a sink receives from various locations in the network, locations are assigned trust values based on the populating trust table in which different factors, such as availability, packet forwarding, etc., are taken into account. Once the trust value is below a specific trust threshold, this location is considered insecure and is avoided when forwarding packets. With the help of some adjustment parameters to dampen oscillation in trust values and absorb the effects of loss of initial packet, this scheme can increase robustness against transient failures and alleviate route infection effects.

Figure 10.7 classifies the secure routing schemes.

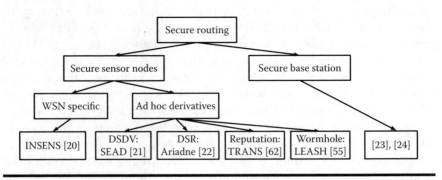

Figure 10.7 Secure routing mechanisms: a taxonomy.

10.7   Broadcast Authentication

Considering the communication patterns in WSNs, as well as local communication that neighboring sensor nodes exchange messages, many-to-one and one-to-many are two commonly used methods. These two cases necessitate broadcast authentication where pair-wise key scheme alone is not enough. However, as mentioned above, traditional public key encryption is undesirable in WSNs, thereby posing a big challenge for broadcast authentication. Fortunately, with the help of a secret key between each sensor node and the base station, the online trusted third parties with more power in both communication and computation, broadcast authentication can be realized in WSNs.

Perrig et al. [14] propose SPINS, a security protocol for bootstrapping keys and broadcast authentication. In this scheme, all sensor nodes intimately trust the base station: at creation time, each node is given a master key that is shared with the base station. All other kinds of keys such as pair-wise key between any two nodes can be derived from the master keys. SPINS consists of two security building blocks: (1) SNEP (Sensor Network Encryption Protocol) and (2) μTESLA. SNEP provides data confidentiality, two-party data authentication, integrity, and freshness via cryptography with a shared counter between the sender and the receiver. μTESLA provides authentication for broadcast data. Instead of using asymmetric keys, μTESLA introduces asymmetry through delayed disclosure of symmetric keys. To broadcast an authenticated packet, the sender computes a MAC (Message Authentication Code) on the packet with a key that is secret at that point in time (will be disclosed after a certain time). When a node receives a packet, it stores the packet in a buffer. At the time of key disclosure, the sender broadcasts the verification key to all receivers. When a node receives the disclosed key, it can easily verify the correctness of the key. If the key is correct, the node can then use it to authenticate the packet stored in its buffer. Figure 10.8 illustrates how it works.

μTESLA is the first work that addresses the broadcast authentication problem in WSNs and provides an efficient approach. However, to bootstrap the authentication broadcast, μTESLA requires the distribution of some initial parameters via unicast communication, which is not efficient,

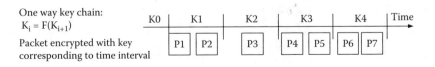

Figure 10.8   Illustration of μTESLA.

especially when there are a large number of sensor nodes in the network. To overcome this problem, a multi-level key chain scheme is extended from μTESLA in which the initial parameters are predetermined and broadcast instead [15]. Specifically, the two-level key chains consist of a high-level key chain and multiple low-level key chains. The low-level key chains are intended for authenticating broadcast messages, while the high-level key chain is used to distribute and authenticate commitments of the low-level key chains. The high-level key chain uses a long enough interval to divide the time line so that it can cover the lifetime of a sensor network without having too many keys. The low-level key chains have short enough intervals so that the delay between the receipt of broadcast messages and the verification of the messages is tolerable. The lifetime of a sensor network is first divided into n (long) intervals of duration and a key is associated with each time interval. Each time interval is further divided into m (short) intervals of duration and a key also associated. When sensor nodes are initialized, all the corresponding information is distributed to the sensor node. Figure 10.9 shows the two levels of key chains. This two-level can be further extended to m-level key chains. The keys in the $(m-1)$-level key chains are used for authenticating data packets. The benefit of having multi-level key chains is that it requires fewer keys in each key chain, or equivalently, shorter duration at each key chain level. Nevertheless, having more levels of key chains does increase the overhead at both the base station and the sensor nodes.

One application of μTESLA is called LEAP (Localized Encryption and Authentication Protocol) [16]. By the observation that different types of messages exchanged between sensor nodes have different security requirements, it is claimed that a single keying mechanism is not suitable for meeting these different security requirements. As a result, four types of keys are established for each sensor node in LEAP: (1) an *individual key* shared with the base station that is generated by a pseudo-random function and pre-loaded into each node prior to its deployment; (2) a *pair-wise key*

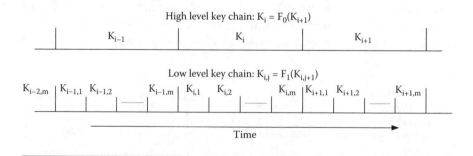

Figure 10.9 Multi-level key chain.

shared with another sensor node that is generated by a challenge-response protocol with the help of pre-loaded master keys; (3) a *cluster key* shared with multiple neighboring nodes and can be generated during the pairwise key establishment phase; and (4) a *group key* shared by all the nodes in the network that is used by the base station for encrypting broadcast-based message and can be pre-loaded before deployment and updated using μTESLA. By establishing different types of keys, LEAP can support in-network processing while at the same time restricting the security impact of a node compromise to the immediate network neighborhood of the compromised node.

Another application of μTESLA can be found in Chen et al. [41].

10.8 Secure Aggregation

Aggregation is one of the main characteristics of sensor networks. In the sensor network with high node density, the result of an individual node does not make much sense compared with the aggregation result from a bunch of nodes in the same region. Aggregation can reduce unnecessary transmission and eliminate redundancy, thus achieving network longevity. However, aggregation needs intermediate nodes to access and modify the data sent by the source, which requires a trust relationship beyond that in traditional end-to-end security mechanisms. Taking this into account, the security mechanism should allow aggregators to access and verify the data, and at the same time prevent them from impersonating other nodes or forging the data, that is, to ensure integrity and confidentiality.

Hu and Evans [37] first address the security problem in information aggregation and present a secure aggregation protocol for WSNs to detect misbehaving sensor nodes by exploiting two main ideas: (1) delayed aggregation and (2) delayed authentication. The first one guarantees for networks where two consecutive nodes are not compromised and the latter one enables authentication keys to be symmetric keys.

By binding a collection of security mechanisms including one-way hash chain, μTESLA, and shared-key establishment protocol, Deng et al. [36] design and implement a dynamic hierarchical WSN to support secure aggregation (in-network processing). With the help of a base station acting as a trusted third party, the mechanism provides authentications in both upstream and downstream directions, and delegates trust to aggregators that are not initially trusted by individual sensor nodes within the network.

When doing aggregation, one challenge comes from compromised nodes. Once a sensor node is compromised, all secret information including cryptography keys is disclosed to the adversary. As a result, the adversary gains full control of the compromised nodes and is able to alter/inject any false data at will to disturb the network. That is, the compromised node can effectively circumvent traditional cryptography and authentication approaches that aim to prevent impersonation. Recently, some effort has been conducted to secure aggregation against compromised nodes' false data injection attacks.

In Zhu et al. [42], an interleaved hop-by-hop authentication scheme is proposed to filter the injected false data. This scheme focuses on nonnumerical data (events, e.g., false alarm) and guarantees that the base station will detect any injected false data packets when no more than a certain threshold number (t) of nodes are compromised. First, this scheme defines the *associated nodes* of a node as the nodes that are ($t + 1$) hops far from it in both directions (uplink and downlink). After network initialization, each sensor node will try to find its associated nodes (called *association discovery*). When an event is triggered, nodes in a cluster (of size t) collaboratively generate a report on this event. Each of the nodes computes two MACs over this event, one using key shared with the base station and the other using a pair-wise key shared with its upper associated node. A cluster header will collect the endorsements from all its cluster nodes as well as itself, synthesize a final report by combining all endorsements, and transmit it to an uplink (toward the BS) neighbor (called *report endorsement*). Upon a node on the path receiving a report, it verifies the authenticity by checking the MAC computed by its lower associated node. If the verification succeeds, it removes that MAC and attaches its own MAC based on the pair-wise key shared with its own upper associated node (called *en-route filter*). When the report finally reaches the base station, the base station verifies the report after receiving it. If it detects that $t + 1$ nodes have endorsed the report correctly, it accepts the report or discards it otherwise (called *base station verification*). In this way, every report generated by a node is authenticated by a node $t + 1$ hops away; that is, the report is authenticated in an interleaved hop-by-hop fashion.

Another authentication-based scheme is called the statistical en-route filtering (SEF) mechanism [43]. Before deployment, each sensor node installs a small number of keys drawn from the global key pool. As a result, each node has a certain probability of possessing one of the keys that some other nodes possess. After deployment, once a stimulus appears, multiple detecting nodes first elect a Center-of-Stimulus (CoS). By summarizing the results, the CoS generates a report on behalf of the group and broadcasts it to all detecting nodes. Each detecting node checks the consistency of the report with its own result and generates a MAC for the report using one of its stored keys if the report passes the check. The CoS collects all the MACs and attaches them to the report before forwarding. When a node on the route receives a report, if it has any key that is used to generate the MACs in the report, it will check the corresponding MAC carried in the report and drops it if they do not match. Otherwise, it passes the report to the next hop. So, as the report is forwarded, each node along the way verifies the correctness of the MACs probabilistically. When the report reaches the sink, the sink that possesses all the keys will check the correctness of every MAC and eliminate any remaining false reports that elude en-route filtering.

The above two schemes both use MAC for authentication, depend on more than one node's agreement on an event to prevent a false alarm, and thus only work for event cases. In Przydatek et al. [35], the authors propose a framework for secure information (numerical data) aggregation using an interactive proof technique. The scheme develops the *aggregate-commit-prove* framework such that the user accepts the data with high probability if the aggregated result is within a desired bound, or rejects the result if it is outside the bound. Different concrete protocols are developed for computing the median, finding the minimum and maximum values, estimating the number of distinct elements, and computing the average of measurement. However, the property of interactive proof introduces communication overhead.

By combining cryptography with tools from other domains (economics and statistics), Ganeriwal et al. [44] propose a reputation-based framework for sensor networks (RFSN) in which nodes maintain reputation for other nodes and use it to evaluate their trustworthiness. This framework employs Bayesian formulation, specifically, a beta reputation system for reputation representation, updates, and integration. A community of trustworthy sensor nodes was formed at runtime based on the behavior of these nodes. This reputation-based framework provides a general approach for not only aggregation (checking outlier), but also routing.

We are also working on secure aggregation. In our approach, we propose an innovative framework for information aggregation by uniquely combining statistical sampling, trust management, and information theory. Specifically, in this framework, data sent from sensors is statistically sampled, evaluated, and measured by aggregators, which will in turn assign reputations to sensors based on relative entropy (or *Kullback-Leibler* distance). By compiling nodes' reputation, the aggregator will forward the aggregation result and its own opinion, indicating its confidence in the result, toward the cluster head. The cluster head will further consolidate multiple aggregation results and corresponding opinions. More importantly, it will incorporate its own opinions toward the aggregators in this process before forwarding the aggregation result to the sink. In addition, our design is based on the powerful and rich Josang's trust model to manage the trust propagation and information theory, which has attributes for a design with strong security performance and robustness.

Figure 10.10 shows the classification of secure aggregation.

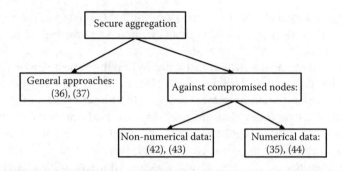

Figure 10.10   Secure aggregation mechanisms: a taxonomy.

10.9   Other Security Issues: Reliability and Fault Tolerance

In sensor networks, several scenarios can cause the network topology to dynamically change, which brings more concerns in reliability and fault tolerance. For example, sometimes, some sensor nodes may be broken or dead due to power-out. Sometimes, sensor nodes might be compromised; after detection, these nodes should be removed from the entire network. Or, an attacker can even remove or disable some sensor nodes from the network. So, sensor networks must be robust and survivable despite individual sensor failures and intermittent connectivity.

Based on the aspect that different information has different levels of importance to the end user and requires different levels of guarantees in delivery, Deb et al. [28] present a hop-by-hop broadcast scheme to ensure information delivery at a desired reliability with minimal cost.

By addressing the problem that conventional end-to-end reliability solutions are not suitable in WSNs, Sankarasubramaniam et al. [29] propose a new, reliable transport scheme, the Event-to-Sink Reliable Transport (ESRT) protocol, to achieve reliable event detection with minimum energy expenditure. The congestion control component in ESRT serves the dual purposes of achieving reliability and conserving energy. The scheme is self-configuring to adapt to the dynamic topology in WSNs. In addition, the algorithms mainly run on the sink, thus relieving the computation burden from the resource-constrained sensor nodes.

Another work [30] focuses on reliability and evaluates different architectural choices on the placement of reliability at different levels of the protocol stack and recommends both the MAC and transport layers. Meanwhile, RMST (Reliable Multi-Segment Transport), a new protocol in conjunction with directed diffusion, is proposed to guarantee delivery even when multiple hops exhibit very high error rates by adding minimal additional control traffic.

For fault tolerance, several schemes have been proposed [31–34]. From different points of view (such as routing mechanism, sensor node faults, feature extraction, etc.), these schemes improve the robustness of the WSN by detecting and correcting faults.

10.10   Conclusion and Open Problems

With the wide applications of WSNs, security issues have received increased attention. The severe resource constraints in the sensor nodes is the main obstacle that affects each aspect of security services in WSNs.

Key management, including key pre-distribution, establishment, and agreement, is the prerequisite for other security services. Probabilistic key sharing is a prevalent scheme although there is no guarantee that the entire network can be completely connected. Some follow-up proposals are deterministic schemes at the price of increasing either communication or computation cost.

The sensor nodes' location can determine the successfulness of geographic routing and is also a primitive for other operations. Such sensitive information necessitates a security mechanism to ensure the integrity from counterfeit.

Routing also faces new challenges. Compared with extensive key management schemes, the works on secure routing are far fewer and are mainly derived from ad hoc networks. However, there are still some differences between ad hoc networks and WSNs, and some attacks are more severe in WSNs. Moreover, it is crucial to consider security issues at the beginning of routing protocol design because sometimes it is unlikely that one will find an effective countermeasure against the attack after completing the design of the protocol.

In-network processing (or aggregation) can reduce transmission redundancy to achieve prolonged lifetime. This mechanism also needs protection, especially when there are compromised nodes in the network.

Another security primitive, authentication, is also very important due to the communication pattern. However, the resource constraints that make nontrivial public key schemes almost infeasible in WSN complicate the broadcast authentication. μTESLA is an effective and efficient scheme for broadcast authentication; even it requires synchronization between sensor nodes and the base station.

Although some efforts have been conducted to secure WSNs from different aspects, there remain a lot of open research issues.

Key establishment. Symmetric cryptography dominates current research because it is a general perception that public key schemes are too expensive for sensor networks. How to explore asymmetric cryptographic mechanisms needs further investigation. More recently, research has shown that public key schemes are indeed feasible.

Huang et al. [18] propose "fast authenticated key establishment protocols" using public key techniques. Different from traditional public key schemes, this approach combines elliptic curve cryptography (ECC) due to the property of small key size and symmetric key operations. Considering the resource constraints in sensor nodes, the scheme reduces the high-cost public key operations (ECC) at the sensor side by replacing them with low-cost symmetric key based operations and puts the cryptographic burden on resource less-constrained devices, the base stations. Meanwhile, authentication of the two identities is still realized on public key certificates to avoid the typical key management problems.

TinyPk [47] provides another example to implement public key technology in WSNs. It shows that sensor networks can employ RSA as the key management scheme for authenticating nodes and distributing key information. Key exchange between nodes is achieved via the Diffie-Hellman scheme.

In the best paper presented at PerCom'05, Eberle et al. show that public key cryptography is viable on constrained platforms even if implemented in software. The authors provide a detailed energy analysis of two public key systems: RSA and ECC. The results show that the energy consumption of such schemes is actually surprisingly small, and can be supported by a single battery throughout the lifetime of a node and hence utilized in wireless sensor networks.

Two more public key cryptosystems in WSNs, Rabin's scheme and Ntru-Encrypt algorithm, are investigated in Kaps et al. [48]. They show that the encryption complexity is on the order of $O(n^2)$. However, developing lightweight asymmetric cryptographic mechanisms for key establishment and digital signatures remains in the early research stages.

Compared with pair-wise keying schemes, group keying protocols receive less attention. Consider the fact that in WSNs, it is common that several neighboring sensor nodes perform the same task to form the clustering topology. In such a case, group key or hybrid keying protocols (pair-wise key *and* group key) might be more efficient.

Routing. How to ensure that the routing protocols work properly even in the presence of compromised adversaries is an essential consideration in routing algorithm design. Additionally, how to effectively prevent different kinds of DOS attacks also challenges routing algorithms design. Furthermore, because key management and routing are intertwined, key distribution needs routing information and routing information should be protected by encryption; integration of routing and keying protocols may bring significant advantages in securing WSNs [25].

Broadcast authentication. Symmetric key (one-way hash function) is most frequently used for encryption and μTESLA is an effective solution. However, it needs time synchronization between the sensor nodes and base station, which might be an issue in some applications and a vulnerability that an attack aims to exploit.

Time synchronization. Synchronization is necessary for both authentication and location measurement; it is an important primitive in WSNs. However, not much work has been done in this domain. One recent work discusses this issue [64].

Infrastructure organization. WSNs usually employ hierarchical clustering built on the sensor nodes to save energy. How to form and maintain the architecture at runtime also brings security concerns. In Law et al. [17], a decentralized key management architecture is presented covering the aspect of key deployment, key refreshment, and key establishment. The main contribution is that it partitions a system into two interoperable security realms: (1) the supervised realm that consists of all supervised clusters and (2) the unsupervised realm that consists of the set of all unsupervised clusters. The supervised nodes are normal sensor nodes; the supervisors are those that are tamper-resistant with higher energy and computational resources, and have wider radio coverage. In the supervised realm, security (confidentiality and authentication) is on the node-to-node communication level within single clusters as well as between different clusters. In the unsupervised realm, the security is on the cluster level; that is, the compromise of a node compromises the security of the whole cluster. And the communication is open and unauthenticated between unsupervised clusters. Different cluster keying protocols are used in different realms to trade off the simplicity and resources.

Agah et al. [65] novelly apply game theory to WSNs and develop cooperative and non-cooperative schemes to guide the cluster formalization.

Intrusion detection. Last but not least, how to detect compromised nodes in WSNs is a vital issue. Because WSNs are often deployed in hostile environments, the sensor nodes are subject to capture by adversaries and the operations are unsupervised. With constrained resources, how to detect the compromised nodes and minimize their effect to ensure that the network operates properly is an intriguing challenge, especially when the compromised nodes already possess all the security information, including secret keys that cause conventional cryptography and, in particular, message authentication code (MAC). One possible solution is to exploit the redundancy in WSNs and use a voting system. However, voting systems themselves are also vulnerable to some attacks, such as bad mouth, etc. This issue influences almost every other security service, such as aggregation, routing, etc., and must be addressed in security design.

References

1. I. Akyildiz, W. Su, Y. Sankarasubramaniam, and E. Cayirci, A survey on sensor networks, *IEEE Communications Magazine*, 40(8):102–114, Aug. 2002.
2. C. Karlof and D. Vagner, Secure routing in wireless sensor networks: attacks and countermeasures, *1st IEEE Int'l Workshop on Sensor Network Protocols and Applications*, May 2003.
3. H. Chan, A. Perrig, and D. Song, Random key pre-distribution schemes for sensor networks, *IEEE Symposium on Research in Security and Privacy*, Orlando, FL, January 2003.
4. L. Eschenaur and V.D. Gligor, A key-management scheme for distributed sensor networks, In *Proceedings of the 9th ACM Conference on Computer and Communications Security (CCS)*, Washington, D.C., November 2002.
5. W. Du, J. Deng, Y.S. Han, and P.K. Varshney, A pair-wise key pre-distribution scheme for wireless sensor networks, *10th ACM Conference on Computer and Communications Security (CCS)*, Washington, D.C., October 2003.
6. R. Blom, An optimal class of symmetric key generation systems, *Advances in Cryptology: Proceedings of EUROCRYPT 84, Lecture Notes in Computer Science*, Springer-Verlag, 209:335–338, 1985.
7. D. Liu and P. Ning, Establishing pair-wise keys in distributed sensor networks, *10th ACM Conference on Computer and Communications Security (CCS)*, Washington, D.C., October 2003.
8. C. Blundo, A.D. Santis, A. Herzberg, S. Kutten, U. Vaccaro, and M. Yung, Perfectly-secure key distribution for dynamic conference, in *Advances in Cryptology — CRYPTO'92*, LNCS 740, pp. 471–486, 1993.
9. W. Du, J. Deng, Y.S. Han, S. Chen, and P. Varshney, A key management scheme for wireless sensor networks using deployment knowledge, in *Proceedings of IEEE Int'l Conference on Computer Communication (INFOCOM)*, 2004.
10. D. Liu and P. Ning, Location-based pair-wise key establishments for static sensor network, *ACM Workshop on Security in Ad Hoc and Sensor Networks (SASN)*, October 2003.
11. S. Zhu, S. Xu, S. Setia, and S. Jajodia, Establishing pair-wise keys for secure communication in ad hoc networks: a probabilistic approach, *11th IEEE Int'l Conference on Network Protocols (ICNP)*, Atlanta, GA, November 2003.
12. R.D. Pietro, L.V. Mancini, and A. Mei, Random key assignment for secure wireless sensor networks, *ACM Workshop on Security of Ad Hoc and Sensor Networks (SASN)*, Fairfax, VA, October 2003.
13. S.A. Camtepe and B. Yener, Combinatorial design of key distribution mechanisms for wireless sensor networks, Technical Report 04-10, Department of Computer Science, Rensselaer Polytechnic Institute, April 12, 2004.
14. A. Perrig, R. Szewczyk, V. Wen, D. Culler, and J.D. Tygar, SPINS: security protocols for sensor networks, *Journal of Wireless Networks*, 8(5):521–534, 2002.
15. D. Liu and P. Ning, Efficient distribution of key chain commitments for broadcast authentication in distributed sensor networks, *10th Annual*

Network and Distributed System Security Symposium, San Diego, CA, February 2003.

16. S. Zhu, S. Seita, and S. Jajodia, LEAP: Efficient security mechanisms for large-scale distributed sensor networks, in *Proc. 10th ACM Conf. on Computer and Communication Security*, Washington, D.C., October 2003.

17. Y.W. Law, R. Corin, S. Etalle, and P.H. Hartel, A formally verified decentralized key management architecture for wireless sensor networks, *4th IFIP TC6/WG6.8 Int. Conf. on Personal Wireless Communications (PWC)*, Wenice, Italy, September 2003.

18. Q. Huang, J. Cukier, H. Kobayashi, B. Liu, and J. Zhang, Fast authenticated key establishment protocols for wireless sensor networks, *WSNA*, 2003.

19. A. Chan, Distributed symmetric key management for mobile ad hoc networks, in *Proceedings of IEEE Int'l Conference on Computer Communication (INFOCOM)*, 2004.

20. J. Deng, R. Han, and S. Mishra, INSENS: intrusion-tolerant routing in wireless sensor networks, Technical Report, CU-CS-939-02, Department of Computer Science, University of Colorado, November 2002.

21. Y. Hu, D. Johnson, and A. Perrig, SEAD: secure efficient distance vector routing for mobile wireless ad hoc networks, *4th IEEE Workshop on Mobile Computing Systems and Applications (WMCSA)*, 2002.

22. Y. Hu, A. Perrig, and D. Johnson, Ariadne: a secure on-demand routing protocol for ad hoc networks, *Proceedings of the 8th Annual Int'l Conference on Mobile Computing and Networking (MobiCom)*, 2002.

23. J. Deng, R. Han, and S. Mishra, A robust and light-weight routing mechanism for wireless sensor networks, *Workshop on Dependability Issues in Wireless Ad Hoc Networks and Sensor Networks (DIWSANS), International Conference on Dependable Systems and Networks*, 2004, Florence, Italy.

24. P. Ganesan, R. Venugopalan, P. Peddabachagari, A. Dean, F. Mueller, and M. Sichitiu, Analyzing and modeling encryption overhead for sensor networks node, *WSNA*, 2003.

25. D.W. Carman, P.S. Kruus, and B.J. Matt, Constraints and approaches for distributed sensor security, NAI Lab Technical Report, #00-010, September 2000.

26. S. Slijepcevic, M. Potkonjak, V. Tsiatsis, S. Zimbeck, and M.B. Srivastava, On communication security in wireless ad hoc sensor networks, *11th IEEE Int'l Workshops on Enabling Technologies: Infrastructure for Collaborative Enterprises (WETICE)*, June 2002, Pittsburgh, PA.

27. L. Lazos and R. Poovendran, Secure broadcast in energy-aware wireless sensor networks (invited paper), *IEEE Int'l Symposium on Advances in Wireless Communications (ISWC)*, Victoria, BC, Canada, September 2002.

28. B. Deb, S. Bhatnagar, and B. Nath, Information assurance in sensor networks, *Proceedings of the 2nd ACM Int'l Conference on Wireless Sensor Networks and Applications (WSNA)*, 2003.

29. Y. Sankarasubramaniam, O.B. Akan, and I.F. Akylidiz, ESRT: event-to-sink reliable transport in wireless sensor networks, *Mobihoc*, 2003.

30. F. Stann and J. Heidemann, RMST: reliable data transport in sensor networks, *1st IEEE Int'l Workshop on Sensor Net Protocols and Applications (SNPA)*, 2003.

31. F. Koushanfar, M. Potkonjak, and A. Vincentellik, Fault tolerance techniques of wireless ad hoc sensor networks, *IEEE Sensors*, 2002.

32. B. Krishnamachari and S. Iyengar, Efficient and fault tolerance feature extraction in wireless sensor networks, *Proceedings of the 2nd Int'l Workshop on Information Processing in Sensor Networks (IPSN)*, Palo Alto, CA, April 2003.

33. J. Deng, R. Han, and S. Mishra, A robust and light-weight routing mechanism for wireless sensor networks, *1st Workshop on Dependability Issues in Wireless Ad Hoc Networks and Sensor Networks (DIWSANS)*, Florence, Italy, 2004.

34. J. Deng, R. Han, and S. Mishra, Enhancing base station security in wireless sensor networks, Technical Report, CU-CS-951-03, Department of Computer Science, University of Colorado, April 2003.

35. B. Przydatek, D. Song, and A. Perrig, SIA: Secure Information Aggregation in sensor networks, *Proceedings of ACM SenSys*, Los Angeles, CA, November 2003.

36. J. Deng, R. Han, and S. Mishra, Security support for in-network processing in wireless sensor networks, *ACM Workshop on Security of Ad Hoc and Sensor Networks (SASN)*, Fairfax, VA, October 2003.

37. L. Hu and D. Evans, Secure aggregation for wireless networks, in *Workshop on Security and Assurance in Ad Hoc Networks.* January 2003.

38. N. Sastry, U. Shankar, and D. Wagner, Secure verification of location claims, *Proceedings of the 2003 ACM Workshop on Wireless Security*, San Diego, CA.

39. T. Park and K. Shin, LiSP: a lightweight security protocol for wireless sensor networks, *ATM Transactions on Embedded Computing Systems*, pp. 634–660, Vol 3, Issue 3, August 2004.

40. A.D. Wood and J.A. Stankovic, Denial of service in sensor networks, *IEEE Computer*, 35(10):54–62, 2002.

41. M. Chen, W. Cui, V. Wen, and A. Woo, Security and deployment issues in a sensor network, *Ninja Project, A Scalable Internet Services Architecture*, Berkeley, http://citeseer.nj.nec.com/chen00security.html, 2000.

42. S. Zhu, S. Setia, S. Jajodia, and P. Ning, An interleaved hop-by-hop authentication scheme for filtering of injected false data in sensor networks, *in IEEE Symposium on Security and Privacy*, pp. 260–272, 2004.

43. F. Ye, H. Luo, S. Lu, and L. Zhang, Statistical en-route filtering of injected false data in sensor networks, in *Proceedings of IEEE Int'l Conference on Computer Communication (INFOCOM)*, 2004.

44. S. Ganeriwal and M.B. Srivastava, Reputation-based framework for high integrity sensor networks, *Proceedings of ACM Security for Ad-Hoc and Sensor Networks (SASN 2004)*.

45. W. Zhang and G. Cao, Group rekeying for filtering false data in sensor networks: a predistribution and local collaboration-based approach, in *Proceedings of IEEE Int'l Conference on Computer Communication (INFOCOM)*, 2005.

46. A. Chadha, Y. Liu, and S. Das, Group key distribution via local collaboration in wireless sensor networks, *IEEE Communication Society Conference on Sensor and Ad Hoc Communications and Networks (SECON), 2005.*

47. S. Cuti, C. Gardiner, C. Lynn, R. Watro, D. Kong, and P. Kruus, TinyPk: securing sensor networks with public key technology, *Proceedings of 2nd ACM Workshop on Security of Ad Hoc and Sensor Networks,* 2004, pp. 59–64.

48. J.P. Kaps, G. Gaubatz, and B. Sunar, Public key cryptography in sensor networks-revisited, in *(ESAS 2004) LNCS 3313,* Heidelberg, Germany, 2004.

49. H. Eberle, V. Gupta, A.S. Wander, N. Gura, and S.C. Shantz, Energy analysis of public-key cryptography for wireless senor networks, *3rd IEEE Int'l Conference on Pervasive Computing and Communications (PERCOM),* 2005.

50. R. Madan, S. Cui, S. Lall, and A. Goldsmith, Cross-layer design for lifetime maximization in interference-limited wireless sensor networks, in *Proceedings of IEEE Int'l Conference on Computer Communication (INFOCOM),* 2005.

51. U.C. Kozat, I. Koutsopoulos, and L. Tassiulas, A framework for cross-layer design of energy-efficient communication with QoS provisioning in multi-hop wireless networks, in *Proceedings of IEEE Int'l Conference on Computer Communication (INFOCOM),* 2004.

52. Z. Li, W. Trappe, Y. Zhang, and B. Nath, Robust statistical methods for securing wireless localization in sensor networks, in *Proceedings of 4th Int'l Conference on Information Processing in Senosr Networks (IPSN),* 2005.

53. L. Lazos and R. Poovendran, SeRLoc: secure range-independent localization for wireless sensor networks, in *Proceedings of the ACM Workshop on Wireless Security (WiSe),* 2005.

54. M. Kuhn, An asymmetric security mechanism for navigation signals, in *Proceedings of the Information Hiding Workshop,* 2004.

55. Y. Hu, A. Perrig, and D. Johnson, Packet leashes: a defense against wormhole attacks in wireless networks, in *Proceedings of IEEE Int'l Conference on Computer Communication (INFOCOM),* 2003.

56. D. Liu, P. Ning, and W. Du, Attack-resistant location estimation in sensor networks, in *Proceedings of the 4th Int'l Conference on Information Processing in Sensor Networks (IPSN),* 2005.

57. S. Brands and D. Chaum, Distance-bounding protocols, in *EUROCRYPT '93,* Vol. 765 of *LNCS.*

58. S. Capkum and J. Hubaux, Secure positioning of wireless devices with application to sensor networks, in *Proceedings of IEEE Int'l Conference on Computer Communication (INFOCOM),* 2005.

59. D. Liu, P. Ning, and W. Du, Detecting malicious beacon nodes for secure location discovery in wireless sensor networks, in *Proceedings of the 25th International Conference on Distributed Computing Systems (ICDCS),* 2005.

60. W. Du, L. Fang, and P. Ning, LAD: localization anomaly detection for wireless sensor networks, in *Proceedings of the 19th IEEE Int'l Parallel and Distributed Processing Symposium (IPDPS),* 2005.

61. L. Lazos, R. Poovendran, and S. Capkum, ROPE: robust position estimation in wireless sensor networks, in *Proceedings of the Int'l Conference on Information Processing in Sensor Networks (IPSN),* 2005.

62. S. Tanachaiwiwat, P. Dave, R. Bhindwale, and A. Helmy, Location-centric isolation of misbehavior and trust routing in energy-constrained sensor networks, in *IEEE Workshop on Energy-Efficient Wireless Communications and Networks (EWCN)*, in conjunction with *IEEE IPCCC*, 2004.

63. J.R. Douceur, The Sybil attack, in *1st Int'l Workshop on Peer-to-Peer Systems (IPTPS)*, 2002.

64. S. Ganeriwal, S. Capkum, C. Han, and M. Srivastava, Secure time synchronization service for sensor networks, in *ACM Workshop on Wireless Security (WISE)*, 2005.

65. A. Agah, S.K. Das, and K. Basu, Security in wireless sensor networks: the game theory approach, in *Proceedings of the ACM Workshop on Wireless Security (WiSe)*, 2004.

SECURE AGGREGATION, LOCATION, AND CROSS-LAYER

V

Chapter 11

Secure In-Network Processing in Sensor Networks

Tassos Dimitriou and Ioannis Krontiris

Contents

11.1 Abstract

In-network processing in large-scale sensor networks has been shown to improve scalability, eliminate information, redundancy, and increase the lifetime of the network. In this chapter we address the challenge of securing in-network processing as it translates to both aggregating sensor node measurements and disseminating commands from aggregators back to individual sensor nodes. We present mechanisms for establishing a secure communication channel between sensor nodes and aggregators, and show how scalability and resiliency to different type of attacks can be achieved. Furthermore, we demonstrate how certain requirements such as low computation/memory overhead and dynamic addition of new nodes in the network can be met. Finally, we elaborate on proposed solutions for resilient aggregation under corrupted measurements and we conclude with some open research directions.

11.2 Introduction

Wireless sensor networks are envisioned to consist of hundreds of thousands of inexpensive nodes, each with some computational and sensing capabilities, operating in an unattended mode and communicating with each other. The purpose of deploying a sensor network is to monitor an area of interest with respect to some physical quantity (e.g., temperature). Information is gathered by the sensors and reported to a point that is referred to as the *base station*. This operation enables a broad range of environmental sensing applications ranging from vehicle tracking to habitat monitoring [1].

Sensor networks are usually characterized by power, processing, and storage limitations. These limitations motivate the design of new network architectures and protocols that try to ensure that the network continues to function and provide data updates for as long as possible, even if sensor nodes operate on batteries and energy can be limited. During data gathering, nodes spend a part of their initial energy on transmitting, receiving, and processing packets. It has been shown [2] that communication takes the largest share, about 70 percent, of the total energy consumption of nodes. Therefore, it is desired that new communication protocols minimize any redundant transmissions and reduce the amount of data relayed in the network, so that the lifetime of sensor nodes is maximized.

Because sensor nodes have limited energy that may be exhausted, sensor networks are densely deployed to deal with connectivity and coverage problems. This causes neighboring nodes to have overlapping sensing regions and generate correlated measurements whenever an event occurs in this overlap. Moreover, sensor nodes are usually deployed randomly, which reinforces the effect of overlapping regions. Each node observes its sensing region independent of its neighbors and sends its measurements to the base station. Therefore, communication overhead can be substantially reduced if raw data sent by nodes can be *combined* to eliminate redundancy and reduce the number of transmissions.

A technique that has been used to deal with these problems is data aggregation (or data fusion) [3–5]. The key idea is to combine highly correlated data coming from different sensors into one packet. This happens at intermediate nodes, called *aggregators*, which compute an aggregated value of all the measurements (e.g., an average or maximum temperature), and then forward only a single packet with the resulted value. The reverse process is called *data dissemination*, in which the network hierarchy is used in the reverse direction to disseminate control messages from the base station towards the aggregators and eventually towards the sensor nodes. For example, as shown in Figure 11.1, in tracking applications the sensor network may need to be used in both modes: first to aggregate sensed data about the movement of the tracked object and then to disseminate commands to nearby sensors to enable further tracking [6].

This and other applications in sensor networks require that sensitive information be delivered to the base station and protected from disclosure to unauthorized third parties. The broadcast nature of the transmission medium, however, makes information more vulnerable than in wired communications. Moreover, due to strict resource constraints, existing network security mechanisms, even those designed for ad hoc networks, are inadequate or inappropriate for this domain, so either they must be adapted or new ones must be created.

All security protocols in sensor networks should satisfy certain requirements so that sensor nodes are able to exchange data securely. The bare minimum consists of providing *confidentiality, authentication, integrity,*

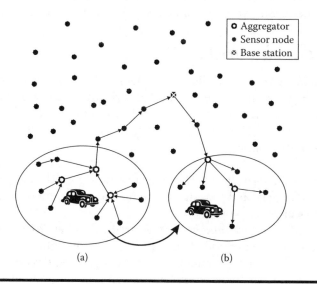

(a)   (b)

Figure 11.1 A tracking application using both (a) aggregation and (b) dissemination.

and *freshness*. However, establishing secure communications between sensor nodes becomes a challenging task, mainly for two reasons: The first is how to bootstrap secure communications between sensor nodes (i.e., how to set up secret keys among them). If we know which nodes will be in the same neighborhood before deployment, keys can be decided *a priori*. Unfortunately, however, most sensor network deployments are random; therefore, such *a priori* knowledge does not exist.

The second reason that makes security in sensor networks difficult is the limited amounts of processing power, storage, bandwidth, and energy resources. Public key algorithms such as RSA are undesirable, as they are computationally expensive. Instead, symmetric encryption/decryption algorithms and hashing functions are between two and four orders of magnitude faster [7] and constitute the basic tools for securing sensor networks communication. Because the resources of a sensor node are very constrained, the key establishment protocols should be lightweight and minimize communication and energy consumption.

In what follows we discuss the threat and trust model for sensor networks and then consider mechanisms to achieve secure in-network processing.

11.3 Threat Model

In sensor network security, an adversary can perform a wide variety of attacks. Not all of them have the same goal or motivations. So, to plan and design better defense systems, we formulate a threat model that distinguishes between two types of attacks: outsider and insider attacks.

11.3.1 Outsider Attacks

In an *outsider attack* (intruder node attack), the attacker node is not an authorized participant of the sensor network. Authentication and encryption techniques prevent such an attacker from gaining any special access to the sensor network. The intruder node can only be used to launch passive attacks, as in the following:

- *Passive eavesdropping*: The attacker eavesdrops (listens) and records (saves) encrypted messages. The messages can then be analyzed to discover secret keys.
- *Denial-of-service attacks*: In its simplest form, an adversary attempts to disrupt the network's operation by broadcasting high-energy signals. In this way, communication between legitimate nodes could be jammed or, even worse, nodes can be energy depleted.
- *Replay attacks*: The attacker captures messages exchanged between legitimate nodes and replays them to change the aggregation results.

11.3.2 Insider Attacks

Perhaps more dangerous from a security point of view is an *insider attack*, where an adversary by physically capturing a node and reading its memory can obtain its key material and forge node messages. Having access to legitimate keys, the attacker can launch several kinds of attacks without easily being detected:

- *False data injection (stealthy attack)*: The attacker injects false aggregation results, which are significantly different from the true results determined by the measured values.
- *Selective reporting*: The attacker stalls the reports of events that do happen, by dropping legitimate packets that pass through the compromised node.

The consequences of these attacks are further magnified in hierarchical networks, where nodes organize themselves in a tree and nodes aggregate results coming from the subtree rooted at them. Therefore, the position of the compromised node can extremely affect the aggregation result. For example, if the attacker compromises a node higher up in the routing hierarchy, she can easily misrepresent the readings of a large portion of the sensor network.

Of course, an adversary cannot have unlimited capabilities. There is some cost associated with capturing, reverse-engineering, and controlling a node. Therefore, we should assume that the adversary can compromise only a limited number of sensor nodes. This fact affects the design of security protocols, as it is easier to offer some protection against a few compromised nodes, but not for the case where a large portion of the network is controlled by the attacker.

11.4 Secure In-Network Processing

In wireless sensor networks, the first step in providing security for data communication is the establishment of shared keys between pairs of nodes so that they can encrypt and authenticate the data exchanged between them. However, caution must be taken so that in-network processing is not hindered by the underlying security protocol. In particular, the following must be taken into account when designing a key management scheme:

1. Data aggregation is possible only if *intermediate* nodes have access to encrypted data so that they can extract measurement values and apply to them aggregation functions. Therefore, nodes that send data packets toward the base station must encrypt them with keys available to the aggregator nodes.

2. Data dissemination implies broadcasting of a message from the aggregator to *all* its group members. If an aggregator shares a different key (or set of keys) with each sensor within its group, then it will have to make multiple transmissions, encrypted each time with a different key, to broadcast a message to all nodes. But transmissions must be kept as low as possible because of their high energy consumption rate.

So, the use of pair-wise shared keys between the nodes and the base station effectively hinders in-network processing. A solution that might satisfy the above requirement would be the use of a key common to all sensor nodes in the network. However, the problem with this approach is that if a single node is compromised, then the security of the entire network is disrupted. Furthermore, refreshing the key becomes too expensive due to communication overhead.

A localized distributed algorithm for key establishment in sensor networks that works well with data aggregation is presented in Dimitriou and Krontiris [8]. A scheme is proposed that utilizes the established keys to provide secure communication between a source node and the base station, while intermediate nodes can access data and perform aggregation. The way this is achieved is by grouping nodes into clusters, where a common key is shared between the nodes of the same cluster. Data is encrypted using that key, so aggregation and dissemination within the cluster is very efficient.

Because information needs to travel between clusters, nodes from different clusters that are in range of each other must share their keys (cluster keys) so that they can encrypt and authenticate messages exchanged between them (see Figure 11.2). This is like creating bonds between clusters to allow secure information flowing. It turns out that, depending on the network density, each node rarely needs to store more than a handful of keys. In addition to that, disclosure of a node's keys allows an attacker to disrupt the communication within this neighborhood; hence, security breaches remain localized.

The above protocol uses clustering only for the key establishment phase. After that phase, communication does not use any hierarchical model. In large sensor networks, however, there is the need to use a hierarchical aggregation model, where aggregation nodes form a tree. As is depicted in Figure 11.3, the main idea is that each aggregator collects information from a subset of nodes (*group* or *cluster*) and then forwards appropriate summaries to other aggregators closer to the base station. When an end user casts a query to the network, required data is obtained more efficiently by this method compared with nonhierarchical approaches. In this case, key establishment must follow a more specific communication scheme, where aggregators collect information from their cluster members and disseminate

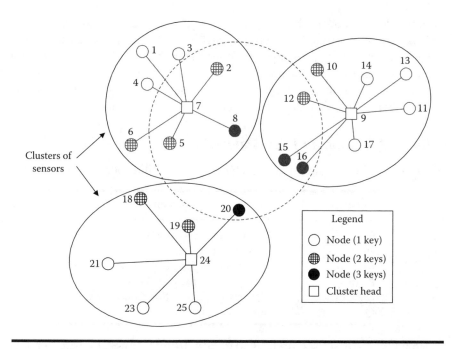

Figure 11.2   An example topology with three clusters. Each node stores its own cluster key as well as the keys of the clusters that are within its communication range. Communication ranges of nodes 25, 17, and 5 are shown.

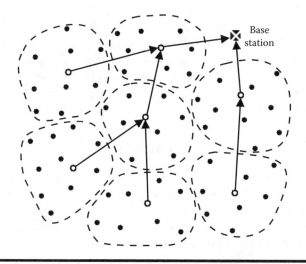

Figure 11.3   A general hierarchical aggregation scheme. The data of each node is gathered by the aggregator that is the representative node in each cluster. Information from aggregators is further aggregated to compute the final result.

commands to them. Based on this scheme, we present an efficient mechanism to establish trust between aggregators and sensor nodes that allows secure in-network processing [9]. Other protocols that focus on securing in-network processing based on different hierarchical organizations and stronger assumptions on the capabilities of sensor nodes are described in Hu and Evans [10] and Deng et al. [11].

11.4.1 Notation and Key Material

Throughout the remainder of the chapter we use the following notation:

Notation	Meaning
S_i	Identifier of sensor node i
A_i	Identifier of aggregator i
M_1, M_2	Concatenation of messages M_1 and M_2
$E_K(M)$	Encryption of message M using key K
$MAC_K(M)$	Message Authentication Code (MAC) of M using key K

Initially, each sensor node is loaded with two keys: (1) a master key K_m, shared among all nodes, and (2) a unique key, denoted K_{S_i} for nodes and K_{A_i} for aggregators[1]. The master key as well as the keys of each sensor/aggregator are known to the base station. For each sensor node S_i, the unique key is computed before deployment as follows:

$$K_{S_i} = F(K_m, S_i)$$

where $F()$ is a secure pseudo-random function. Notice that an adversary upon compromising node S_i cannot recover the master key K_m from the key K_{S_i} because of the one-wayness of $F()$. Hence, the rest of the network remains secured.

Correspondingly, for each aggregator A_j, the unique key K_{A_j} is derived from

$$K_{A_j} = F(K_m, A_j)$$

We see in Section 11.4.2 that K_m is kept by A_j (and by the S_i's) for as long it is necessary to establish secret keys with the nodes belonging in the A_j's group. Then it is *deleted* from the memory of the aggregator as well as of the rest of the nodes so as to eliminate the possibility of being retrieved by an adversary who has physically captured a node.

---

[1] We do not imply by this that the aggregators are known in advance. Once nodes determine their roles through the use of some clustering protocol, then we can talk about the keys K_{S_i} and K_{A_j}.

In what follows, although not shown, we assume the use of different keys for different cryptographic operations; this prevents potential interactions between the operations that might introduce weaknesses in a security protocol. Therefore, we use independent keys for the encryption and authentication operations, K_{encr} and K_{MAC}, respectively, which are derived from each unique key K_{S_i} through another application of the pseudorandom function F (e.g., $K_{encr} = F(K_{S_i}, 0)$ and $K_{MAC} = F(K_{S_i}, 1)$).

11.4.2 Key Establishment for Secure Aggregation

As discussed in the previous section, each sensor node is preloaded with the key K_{S_i}, which is the result of the application of a secure pseudorandom function on the master key K_m. For a node to communicate with its aggregator, this key must be available in both sides. So, the node must first inform the aggregator about its key in a secure way. This can happen by sending an appropriate "hello" message M_{Hello}:

$$S_i \rightarrow A_j : M_{Hello}, MAC_{K_{S_i}}(M_{Hello})$$

where the M_{Hello} consists of the sensor's id S_i and a nonce N_{S_i} computed by the sensor.

Having received this message, the aggregator A_j is now able to compute K_{S_i} using the master key K_m and authenticate it by checking the MAC. If the MAC verifies, the sensor node is included in the cluster and the aggregator sensor stores all relevant information (such as K_{S_i}, nonce identifier — to avoid replay attacks, etc.). Additionally, the aggregator can send back a reply to acknowledge the inclusion in the cluster (also MACed with K_{S_i}).

The same procedure is repeated for every node and its corresponding aggregator in an initial phase after the deployment of the network. This phase is secure as long as the adversary does not know the master key. Therefore, we must assume that the phase is too short in time for an adversary to capture a node and retrieve the master key K_m (see also [8,12] for a similar assumption).

The memory requirements of this phase are determined by the number of keys each aggregator has to store (i.e., the number of nodes in each cluster). Figure 11.4 shows the average number of keys each aggregator has to remember as a function of the number of aggregators in the network. As expected, the more aggregators in the network, the more clusters that will be formed and, therefore, the less the average number of sensor nodes in each cluster.

However, because we consider random topologies, it may be the case that one cluster has a lot more members than another. Figure 11.5 shows the maximum number of keys found in aggregators after 1000 experiments

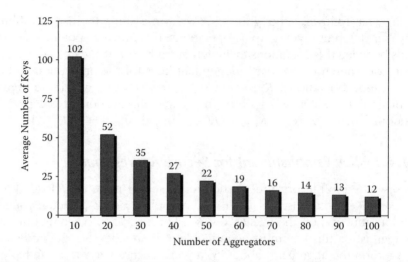

Figure 11.4 Average number of keys an aggregator must store in a random network of 1000 nodes.

in random networks of 1000 sensor nodes. In all the experiments, 50 aggregators were randomly selected and each sensor node joined the cluster of the closest aggregator. The figure shows the distribution of the *maximum* number of keys found in any aggregator, which is within reach of current sensor node memories.

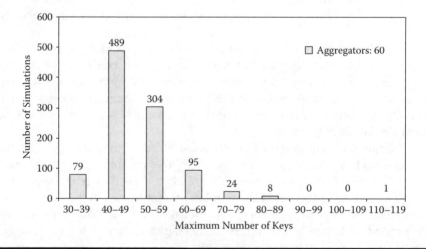

Figure 11.5 Maximum number of keys an aggregator must store in a random network of 1000 nodes. The figure shows that in most cases (432 experiments out of 1000), the maximum number of keys found in any aggregator was between 50 and 59.

After the completion of this phase, the master key is *deleted* by the memory of every node. Because the security of our protocol depends on the deletion of the master from the memory of the sensor nodes, we should take care that the deletion is unrecoverable, for example, by overwriting the master key (in practice several times). We have now established a secure channel between the aggregators and the sensor nodes.

11.4.3   Key Establishment for Secure Dissemination

Having secured the communication of sensor nodes toward their aggregators, we now need to secure the reverse procedure: the dissemination of messages/commands *from* aggregators *to* individual sensor nodes. One simple but inefficient solution would be to send a separate unicast message to each member of the group encrypted and authenticated using the key shared between the aggregator and the specific node. This would result in several transmissions of the same message, wasting valuable energy resources.

The solution to this problem is for the aggregator to construct and propagate a *group key* to its members during the initial phase after deployment. In this way, dissemination can take place later by a single broadcast of the message encrypted by that key. So, each aggregator A_i constructs and sends a group key G_{K_i} to each sensor S_j that belongs to its group:

$$A_i : \quad c = E_{K_{encr}}(\text{"Group key"}, A_i, G_{K_i}),$$
$$\sigma = MAC_{K_{Mac}}(c)$$
$$A_i \to S_j : \quad A_i, c, \sigma$$

The group key is encrypted and authenticated each time using the sensor's private key K_{S_i}, which is available to the aggregator as described in the previous section. First, the aggregator encrypts the group key using the key K_{encr} derived from K_{S_i} and then creates a MAC σ of the resulting ciphertext c.

Once the aggregators settle a group key G_{K_i} with their group members, they can broadcast commands encrypted and authenticated using G_{K_i}. This is sufficient to secure communication from an outsider who does not hold the group key. However, an insider adversary who has captured a group node and retrieved G_{K_i} can impersonate the aggregator and send forged messages to nodes in the same group. Therefore, we need to further secure the dissemination process.

11.4.3.1   Defending against Impersonation Attacks

We must enhance the security protocol described above toward off insider attacks that impersonate the aggregator and try to disseminate messages to the group. The solution is simple and elegant: whenever an aggregator A_i

has a new command to disseminate to the nodes, it attaches to it the *next* key, K_l, from a *one-way key chain*, as follows:

$$A_i : \quad c = E_{G_{K_i}}(\text{"Command"}, A_i, K_l),$$
$$\sigma = MAC_{G_{K_i}}(c)$$
$$A_i \rightarrow Group: \quad A_i, c, \sigma$$

Thus, the *l*-th command sent by the aggregator to the group nodes contains the *l*-th commitment of the hash chain and is encrypted and authenticated using keys derived from G_{K_i}.

One-way key chains are a widely used cryptographic primitive. To generate a chain of length *n*, we randomly pick the last element of the chain K_n. Then, each element of the chain is generated by repeatedly applying a one-way function F until the K_0 element, which is the commitment to the whole chain. We then reveal the elements of the chain in reverse order:

$$K_0, K_1, \ldots, K_l, \ldots, K_{n-1}, K_n$$

If we know that K_{l-1} is part of the chain, we can verify that K_l is also part of the chain by checking that $K_{l-1} = F(K_l)$. Therefore, a node receiving a command encrypted with the group key can verify its authenticity by checking whether the new commitment K_l generates the previous one through the application of F. When this is the case, it replaces the old commitment K_{l-1} with the new one in its memory and accepts the command as authentic. Otherwise it rejects it.

One issue with the one-way key chain is that it limits the length of the trust "delegated" by the aggregator. This is because the length of the chain determines the number of packets that the aggregator can send to its sensor group members. That is, for a chain of length *n*, the aggregator can send at most $n - 1$ separate commands at the nodes. However, an aggregator can renew its key chain as follows: before the aggregator uses the last commitment, it creates a new hash chain $K_0', K_1', \ldots, K_{n-1}', K_n'$ and broadcasts a "renew hash chain" command that contains the new commitment K_0' authenticated with the last unused key of the old chain. This essentially provides the connection between the two chains and the group nodes will be able to authenticate commands as before.

Thus far we have limited the possibility of an impersonation attack, but we have not eliminated it. There is still a scenario wherein an adversary can *jam* communications to a sensor S_j so that it misses the last *k* commands and hence commitments. Then it introduces new commands by "recycling" the unused commitments. It will be impossible for sensor S_j to notice the faked commands as the ordering of commitments is followed.

To defend against an attack such as this, we can assume that sensors are loosely time synchronized and commands are issued only at regular time intervals. That means a packet broadcast from the aggregator at time slot t contains commitment K_t. If a sensor node does not receive anything within the next d slots and the last commitment was K_t, it will expect to see the commitment K_{t+d} that accounts for d missing commands. In this way it cannot be misled and authenticate unused commitments.

Finally, we must note that an implicit defense against impersonation attacks that not only eliminates but also helps detect these types of attacks is to have the sensor nodes respond to the aggregator using the shared key K_{S_i}. In this manner, even if an attacker has issued false commands, it will not be able to use the information sent by the sensor nodes. Additionally, if the aggregator receives responses to commands that it has not issued, it will become aware that an attacker has compromised some nodes in the cluster and eventually take corrective actions.

11.4.4 Adding New Nodes to the Network

This section addresses the problem of refreshing the network, as sensors usually have limited lifetimes and usually die of energy depletion. Each new node S_i' comes equipped with a unique key, $K_{S_i'}$, which is derived from a new master key K_m' in the way we have already described. The new master key is also forwarded to all the aggregator nodes from the base station, using the key shared between the base station and the aggregator to create a secure channel.

Every new node transmits a Hello message to its neighbors indicating its will to become a member of an existing group. The message contains its identifier and a nonce:

$$S_i' \rightarrow A_j : M_{Hello}, MAC_{K_{S_i'}}(M_{Hello})$$

In the same manner as described in Section 11.4.2, the aggregator computes $K_{S_i'}$ and authenticates the Hello message. If all checks out, the new node is added to the cluster and the aggregator creates a new group key that broadcasts to the group.

11.5 Resilient Aggregation

Thus far we have described a key management scheme that allows aggregators to authenticate sensor nodes, and vice versa. As a result, a malicious node injected in the network cannot authenticate itself and disrupt in-network processing. If, on the other hand, an adversary physically captures an existing node, the compromised node will still authenticate itself to the aggregator and may start sending false readings, in an effort to affect the aggregation result. It is difficult to detect such an attack because that would require application (semantics)-specific knowledge.

What can be done in this case is to make aggregation functions more resilient to some corrupted measurements. This means that a few corrupted measurements should not cause large errors in the computed aggregate, under the assumption that only a very small fraction of nodes can be compromised. In research work done in this direction [13], it is proposed that instead of aggregating measurements by computing the sum or the average, more resilient functions can be used. For example, the median or the 5 percent-trimmed mean, where the highest and lowest 5 percent of the sensor readings are ignored, can be fairly robust for large network sizes and high node densities.

Another approach is described in Przydatek et al. [14] that is termed the aggregate-commit-prove technique and used to authenticate the results sent by aggregators in the base station. That is, even if an attacker captures the aggregator and forces it to produce false results (within a bound), the attack can still be detected with high probability. First, the aggregator computes the aggregation result of data collected by sensor nodes. In the commit phase, the aggregator commits to the collected data, which ensures that the data collected was actually used. For the commitment, a Merkle hash-tree is used [15], where the leaves hold the collected data and each internal node computes the hash value of the concatenation of the two child nodes. The root (aggregator) computes the final hash value, which is the commitment. In the final phase, the aggregator sends the aggregated data and the commitment to the base station. The aggregator then proves to the base station that the data is correct using an interactive proof. To do that, the base station does the aggregation on some randomly chosen samples to check if the aggregated result sent by the aggregator is valid.

The proposals presented in this section are some first steps toward a general need to fortify sensor network security mechanisms with self-awareness and self-healing so that they can autonomously recognize and respond to a security threat. This requires that nodes have knowledge about the network's state (or, more realistically, the state of neighboring nodes) and monitor the network for abnormal behaviors of sensor nodes or data traffic. To characterize normal and malicious behavior, appropriate rules must be generated, based on statistics, induction, and deduction. Once the network is aware that an intrusion has taken place and has detected the compromised area, appropriate action must be taken. The first action is to cut off the intruder as much as possible and isolate the compromised nodes. After that, proper operation of the network must be restored. The challenge here is that all these steps must be taken autonomously, that is, without human intervention.

11.6 Conclusions

In this chapter we have presented mechanisms for secure in-network processing in sensor networks. We first elaborated on the general requirements that such an operation must satisfy and then described an efficient security protocol for a general hierarchical aggregation/dissemination scheme. This protocol provides a key management scheme for the establishment of secure channels between nodes and aggregators, as well as support for the addition of new nodes to the network, a critical requirement in sensor networks as sensors have limited energy and thus limited life expectancy.

Performance evaluation shows that secure in-network processing can be achieved with very realistic memory and processing requirements. Our protocol does not depend on the existence of location information or on the underlying routing protocol. There is also no need to burden the base station with any additional computation load. Secure in-network processing can be realized as a distributed service that is scalable and adaptable.

An extension to the proposed protocol would be to consider the case where aggregators must delegate their authorities. This seems to be the case because aggregators consume more energy than regular sensor nodes and their resources may be drained faster. An immediate consequence of this would be to consider mechanisms where new nodes will take the role of the aggregator or clusters will have to be reorganized so that some form of load balancing is achieved.

As another research direction, note that the mechanisms presented here can be part of a more general protocol that tries to defend against a broad range of attacks. Of course this is a huge research challenge, where sensor networks must be reinforced with adaptive security architectures that can monitor the network, recognize a security threat, and respond either by preventing the intruder or by isolating the damage and restoring the network's normal operation. In a dynamic communication environment of thousands of nodes, such mechanisms must not hinder other network processes, but rather coexist with them and protect them.

References

1. C.-Y. Chong and S.P. Kumar, Sensor Networks: Evolution, Opportunities, and Challenges, in *Proceedings of IEEE*, Vol. 91, No. 8 (2003) pp. 1247–1256.
2. A. Perrig, R. Szewczyk, V. Wen, D. Culler, and J.D. Tygar, SPINS: Security Protocols for Sensor Networks, in *Proceedings of 7th ACM Mobile Computing and Networks* (2001).
3. C. Intanagonwiwat, R. Govindan, and D. Estrin, Directed Diffusion: A Scalable and Robust Communication Paradigm for Sensor Networks, in *Proceedings of 6th ACM/IEEE Mobicom Conference* (2000).

4. B. Krishnamachari, D. Estrin, and S. Wicker, The Impact of Data Aggregation in Wireless Sensor Networks, in *Proceeding of International Workshop on Distributed Event-Based Systems* (2002).

5. S.R. Madden, R. Szewczyk, M.J. Franklin, and D. Culler, Supporting Aggregate Queries over Ad-Hoc Wireless Sensor Networks, in *Proceedings of Workshop on Mobile Computing Systems and Applications* (2002).

6. Y. Xu, Energy-Aware Object Tracking Sensor Networks, in *Proceedings of International Conference on Distributed Computing Systems, 2003 Doctoral Symposium* (2003).

7. D. Carman, P. Kruus, and B.J. Matt, Constraints and Approaches for Distributed Sensor Network Security, Tech. Rep. 00-010, NAI Labs, 2000.

8. T. Dimitriou and I. Krontiris, A Localized, Distributed Protocol for Secure Information Exchange in Sensor Networks, in *Proceedings of the 5th IEEE International Workshop on Algorithms for Wireless, Mobile, Ad Hoc and Sensor Networks* (2005).

9. T. Dimitriou and D. Foteinakis, Secure In-Network Processing in Sensor Networks, in *Proceedings of First Workshop on Broadband Advanced Sensor Networks (IEEE BASENETS)* (2004).

10. L. Hu and D. Evans, Secure Aggregation for Wireless Networks, in *Proceedings of the Symposium on Applications and the Internet Workshops* (2003).

11. J. Deng, R. Han, and S. Mishra, Security Support for In-Network Processing in Wireless Sensor Networks, in *Proceedings of the 1st ACM Workshop on Security of Ad Hoc and Sensor Networks* (2003).

12. S. Zhu, S. Setia, and S. Jajodia, LEAP: Efficient Security Mechanisms for Large-Scale Distributed Sensor Networks, in *Proceedings of the 10th ACM Conference on Computer and Communication Security* (2003).

13. D. Wagner, Resilient Aggregation in Sensor Networks, in *ACM Workshop on Security of Ad Hoc and Sensor Networks (SASN'04)*, (2004).

14. B. Przydatek, D. Song, and A. Perrig, SIA: Secure Information Aggregation in Sensor Networks, in *Proceedings of the ACM SenSys* (2003).

15. R.C. Merkle, Protocols for Public Key Cryptosystems, in *Proceedings of the IEEE Symposium on Research in Security and Privacy* (1980).

Chapter 12

Secure Localization in Sensor Networks

Karthikeyan Ravichandran and Krishna M. Sivalingam

Contents

12.1 Abstract

In this chapter we present a survey of "secure localization" systems for wireless sensor networks. Location information is an important component of sensor data in many wireless sensor networks. However, much of the earlier work in this area did not consider security aspects. Recently, research on location estimation systems that deal with a hostile environment has attracted much attention. We present a comprehensive survey of the several proposed secure localization techniques. We provide a taxonomy of the schemes, along with the important details of each scheme. We also present a comparison study of the main characteristics of these schemes.

12.2 Introduction

Location information of wireless devices has been found useful for various applications. Location information can also be useful in communication protocols such as routing [1,2] and also in service discovery in a new environment. Accurate locationing becomes all the more important in sensor networks as location information helps to extract maximum knowledge from the data obtained from sensor measurements. For example, a collection of sensors forming an intrusion detection system will be useful only if it can pinpoint the location of intrusion. Localization, the process of location estimation of wireless devices, has been a fairly well researched topic culminating in the development of systems such as RADAR [3], Cricket [4], Active Badge [5], and many other techniques proposed in [6–8].

Almost all the localization systems mentioned above operate in a non-adversarial setting. We define an adversary as an attacker who tries to sabotage the process of localization by either forcing the system to make an incorrect estimate of the location (location spoofing) [9] or by making localization impossible (denial of service). Adversaries can also employ different attacks, such as replay, wormhole, etc. Almost all the existing localization systems tend to fail in the presence of adversaries. This is due to the reason that in the absence of adequate precautions, the adversary could easily modify the distance estimates of the nodes. Deployment of these schemes in highly critical scenarios such as reconnaissance and intrusion detection will lead to disastrous consequences. Even in a non-mission-critical setting such as routing in a mobile ad hoc network, false location information can lead to a significant performance degradation of the routing protocol. Thus, it is essential that location estimation must be secure at all times in these applications. This provides us with significant motivation to study techniques that provide location estimation in a secure manner. This process of accurately estimating the position of a node in a hostile environment is called *secure localization.* Secure localization has only recently gained research attention, but much more must be done, and we intend to capture the state-of-the-art in this survey.

A related problem is the *location verification* problem, wherein the system must securely validate the location claims of a user. This finds applications in scenarios where a user gets certain privileges based on his location. In this case, the mere fact that the person has presence in the location serves to signify his privileges to the system. For example, attendees at a conference in a particular area could be given access to information that is normally denied to regular users of the system.

The remainder of this chapter is organized as follows. In Section 12.2, we provide a background of localization techniques and a model of an adversary who might try to disrupt the process of localization. In Section 12.3, we discuss the various secure localization techniques and classify them based on their characteristics. We also provide a comparison of the existing schemes that discusses the limitations and deficiencies of the existing systems. We present our conclusions in Section 12.4.

12.3   Background

This section presents an introduction to localization systems and the different types of attacks on localization systems.

12.3.1   Localization

We present a brief introduction to localization systems that can serve as a foundation for our work on secure localization. Localization techniques typically use distance estimates and a triangulation process to determine the position. However, not all systems use distance estimates. Distance and position estimating systems can be broadly divided into two categories:

1. Ranging based
2. Range free

Ranging-based systems rely on accurately estimating the distance between nodes by analyzing the physical properties of signal transmission between the two nodes. Ranging can be done by estimating one of the following properties:

- *Received Signal Strength-based techniques* use the received signal strength characteristics to determine the distance between two nodes. Any signal transmitted will suffer from attenuation due to path loss. By knowing the transmitted signal strength T and by measuring the received signal strength R, the distance d between any nodes is calculated as $d = f(T, R)$. The function f is the path loss model of the wireless channel and can be estimated by channel estimation techniques.

■ *Time of Arrival (ToA)-based systems* measure the round-trip signal propagation time between the two nodes performing distance estimation. A node sends a message to a receiver, which then replays the same message. The round-trip propagation time is recorded as the time difference between signal transmission and reception of the replayed message. From the knowledge of round-trip time, the distance can be measured as a function of this round-trip time and velocity in the medium.

■ *Time Difference of Arrival (TDoA)-based systems* operate on the same principle as time of arrival based systems. Here, the time of message reception at different points of a transmitted message is recorded. The distance can be measured as a function of the time difference between the different recordings (at different points). In two-dimensional systems, we need at least three receivers of the signal and the location can be determined by the point of intersection of the hyperbolas based on the time difference of arrival between signal reception at multiple receivers.

■ *Angle of Arrival (AOA) techniques* use the angle of signal reception at a node to estimate the direction of the transmitting node. This can be coupled with the distance estimate to provide more accurate node locations.

The nodes that already know their position and hence help another node to determine its position are called *beacons*. In a two-dimensional space, once the distance of the node to at least three beacons is known, the position of the node can be found by triangulation. This is the basic principle behind the range-based techniques. In the above-mentioned techniques of distance estimation, the ToA and TDoA techniques can typically be performed with RF and ultrasound. However, ultrasound-based ranging is more practical because the signal travels at the speed of sound, which is significantly lower than the speed of light. Thus, the effect of finite processing time to process the signal will be felt less in an ultrasound-based system.

Some of the locationing systems that have been developed include the commercially available outdoor Global Positioning System (GPS) with location accuracy on the order of a few meters. Indoor localization systems include RADAR [3], with an accuracy of one meter; Horus [10], with an accuracy of one meter; and Cricket [4], with an accuracy of a few centimeters. The above-mentioned systems were developed mainly for a wireless LAN scenario. Savvides et al. [11] developed a localization system for wireless sensor networks called Ahlos that can accurately measure locations on the order of a few centimeters. The accuracy of localization systems that use signal strength has been extensively studied in Elnahrawy et al. [12] and gives insight into the limitations of RF-based locationing systems.

Range-free systems do not employ range estimates to calculate node positions. Instead, they use information about the location of existing nodes to determine the position of the node of interest. In the simplest of schemes called centroid [6], the authors calculate the position of the node as the centroid of the triangle formed by its three neighbors. Another scheme, called APIT [7], uses an approximation of the point in a triangle test to determine node positions. In another scheme called DV-based positioning [8], the number of hops of the node from a reference node is determined. From the number of hops, an estimate of node distance is obtained and the node position is calculated by trilateration. Range-free systems have less complexity compared to range-based systems, but also have lower accuracy. Hence, range-based schemes are preferred over range-free schemes when high accuracy is desired.

12.3.2 Attacks on Localization

To analyze the attacks that are possible on localization schemes, we need to define the characteristics (strengths and limitations) of the attacker. The attacker can be a node external to the system that does not possess the cryptographic quantities necessary to associate with the system. We call such an attacker a *malicious node*. On the other hand, the attacker could have obtained a handle over the cryptographic entities of one of the nodes in the system and hence will appear as a legitimate node to other nodes of the system. Such an attacker is a *compromised node*. Compromised nodes appear as part of the network and are thus inherently more problematic than malicious nodes. Nevertheless, both kinds of attackers are common, and any secure localization technique must be able to combat both of them effectively.

In range-based techniques, the attackers strive to disrupt the distance estimation process. If the distance estimate from the node to the beacons is false, the nodes will not be able to accurately determine their location using trilateration. In range-free systems, the attackers try to propagate false information within the network regarding position estimates. This will seriously undermine the position estimation technique. For example, in centroid [6], if the location of a node forming the triangle is falsely advertised, then the centroid and hence the calculated node position will be erroneous.

As mentioned, all existing localization schemes are vulnerable to attacks and hence location accuracy cannot be guaranteed. We now look at different types of attacks against localization systems and the vulnerability of the existing schemes to these attacks.

12.3.2.1 Distance Modification

In this kind of attack, the attacker introduces error in the node's distance estimate to the beacons. If signal strength is used for ranging, the attacker can jam the transmission of the sender and replay the same message with a significantly higher/lower power level, thus decreasing/increasing the node–beacon distance. In the case of Time of Arrival (ToA) and Time Difference of Arrival (TDoA) based schemes, two scenarios are possible. When ultrasound is used for ranging, the attacker can jam the sender's signal and choose to replay it back to the receiver. In replaying, the attacker can use a radio link that is significantly faster than ultrasound to reproduce the signal to the receiver. In this case, the receiver will estimate a distance that is shorter than the actual distance. On the other hand, the attacker can just choose to delay the message transmission, which will result in a longer distance estimate. Thus, when ultrasound is used for ranging, both distance enlargement and distance reduction attacks are possible. When radio frequency (RF) is used for ranging, the attacker can jam the sender's signal and replay the message at a later time. He cannot increase the speed of the signal, as RF waves propagate at the speed of light. Hence, when RF is used for ToA or TDoA measurements, the attacker can only increase the distance between the two nodes; he can never make the nodes appear closer than they are. It must be noted that distance modification attacks are frequently employed against range-based systems. Because range-free systems do not employ ranging, they are not vulnerable to distance modification attacks.

12.3.2.2 Wormhole Attack

Whenever an attacker controls more than one node and can establish a communication link between them, a wormhole attack is theoretically possible. In a wormhole attack, the attacker replays messages that it hears from one end of a wormhole through the other end of the wormhole. Thus, a node can hear messages that it would not normally hear (due to communication range constraints) through a wormhole. This will potentially feed false information to the nodes regarding the location of its neighbors. In our case, a node may hear location advertisements of a non-neighbor through a wormhole. The node will consider the advertising node as a neighbor (within its communication range) and hence position estimation will turn out to be incorrect. A typical wormhole scenario is shown in Figure 12.1. Nodes 4 and 5 form a wormhole link and they transmit messages from node 1 to node 6, and vice versa. Hence, nodes 1 and 6 assume that they are within the communication range of each other and may calculate wrong positions in the absence of adequate defenses. A wormhole attack is one of the most popular attacks in wireless networks and is applicable in other

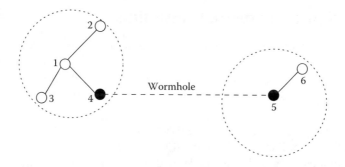

Figure 12.1 Example of a wormhole attack, where nodes 4 and 5 are malicious.

scenarios, such as secure routing. Wormhole attacks can be detected by *packet leashes* [13]. With regard to localization systems, wormhole attacks are more common against range-free systems than against range-based systems. This is due to the fact that range-based systems typically employ Time of Arrival (ToA) or Time Difference of Arrival (TDoA) for ranging, and the existence of wormhole simply signifies an extended communication link between the nodes concerned. Because these systems also employ trilateration, wormholes must be established between all three beacons that participate in trilateration.

12.3.2.3 Replay Attack

In replay attacks, the attacker usually transmits legitimate messages transmitted by the nodes at a later point in time. This enables the propagation of stale data in the system. For example, a node might have changed its position (due to mobility) after its previous location advertisement and hence location advertisements are no longer correct. Replay of these old messages by the attacker might lead to incorrect position determination. It must be noted that this is a relatively simple attack for the attacker compared to wormhole because it needs only one malicious attacker and it does not need a separate communication network between attackers.

12.3.2.4 Sybil Attack

The Sybil attack [14] refers to the process by which a node takes on different identities at different points in time. When a Sybil attack is employed against localization, a single malicious/compromised node can assume the identity of multiple nodes and thus the system sees an increased number of attackers. Thus, when a scheme is susceptible to Sybil attack, the system is vulnerable even to a small number of malicious nodes. Sybil attacks in sensor networks have been extensively studied in Newsome et al. [15].

12.4 Sensor Network Constraints

Sensor networks have several unique constraints that necessitate solutions that optimize these parameters [16]. Hence, a secure localization solution for a general wireless network may not necessarily be the optimum solution for sensor networks. Some constraints unique to sensor networks are:

- *Power:* Sensor nodes are heavily constrained by power because of their form factor and due to the fact that they run on stored energy devices such as batteries. Hence, it is of paramount importance to conserve power expended in these devices. Any localization scheme should take this fact into account and hence use power as efficiently as possible. Another important factor to consider is the distribution of power dissipation. In some cases, lack of power in some of the nodes might result in complete disruption of network services as a whole. Hence, we would ideally want all nodes in the network to dissipate a uniform amount of power. This would be possible only if all the tasks, including localization, are distributed uniformly over the network. This also provides motivation for considering distributed localization algorithms rather than centralized algorithms.

- *Memory:* Memory is a very scarce commodity in sensor networks as the addition of memory results in higher cost and power consumption. This would hamper the deployment of thousands of these nodes. Hence, it would always be better to choose techniques that will minimize memory usage within the sensor node.

- *Computational capability:* A typical desktop processor runs on the order of gigahertz (GHz). In comparison, the processors that typically reside in a sensor node are restricted to a few hundred megahertz (MHz). Thus, highly computation-intensive algorithms are not suitable for processing by sensor nodes. The computation performed at a sensor node must be kept to the minimum. Also, sensor nodes typically perform many other important tasks in addition to localization. Hence, computational complexity will be another parameter that must be considered when considering secure localization algorithms for wireless sensor networks.

12.5 Secure Localization Schemes

Securing position estimation is currently a relatively new area of study in the research community. In this section we attempt to summarize the currently known secure localization techniques. Figure 12.2 provides a classification of the different schemes according to the technique used for secure positioning. The mechanism presented in Li et al. [9] belongs to the class of statistical filtering-based algorithms. The other algorithms [17–19] do not

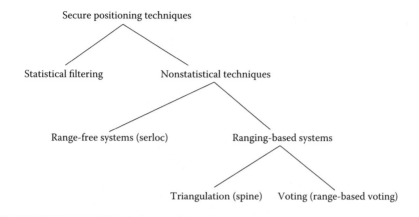

Figure 12.2 Classification of secure localization schemes.

employ filtering techniques. Among these algorithms, Serloc [18] is range independent, whereas SPINE [17] and voting based [19] employ ranging for secure localization. We also include a discussion of the location verification problem in this section.

12.5.1 Verifiable Multilateration Based on Distance Bounding

Distance bounding [20] provides a technique to establish a lower bound on the distance between two nodes. Hence, the two nodes cannot estimate their distance to be smaller than their actual distance. However, distance bounding does not provide an absolute estimate of the distance between two nodes. The attacker may still be able to enlarge the distance between any two nodes. Capkun and Hubaux [17] propose a solution in which the position of a node is calculated when it falls within a triangle formed by three other beacon nodes. The beacons are called verifiers and all the verifiers perform distance bounding with u, the node that wants to determine its position. Distance bounding requires a shared secret key between the node and the verifier. Distance bounding in this case involves the calculation of the round-trip propagation time of a signal transmitted between the node and the verifier. The signal transmitted by the verifier is processed by the node and sent back to the verifier. The time difference between the message transmission by the verifier and receipt by the node is given by

$$t = rtt_{uv} + t_p(u)$$

where rtt_{uv} is the round-trip signal propagation time between u and v, and $t_p(u)$ is the processing time at u. The actual distance can be bounded by $d = rtt_{uv} \times c$, where c is the speed of RF transmission. Hence, we would

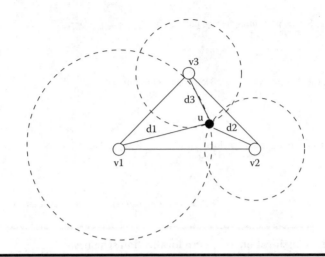

Figure 12.3   Position estimation based on trilateration in SPINE.

like the processing time at u, $t_p(u)$, to be as low as possible. To achieve such a small processing time on the order of nanoseconds, the processing will be a simple operation like XOR. To prevent replay of messages by an attacker, encrypted random nonces are used. The precise details of the algorithm can be found in Capkun and Hubaux [17].

Once distance bounding is performed between the node and the verifiers, the verifiers utilize trilateration to determine the position of node u. In Figure 12.3, the verifiers $v1$, $v2$, and $v3$ bound the distance to node u as $d1$, $d2$, and $d3$, respectively. The attack-resistant property of this scheme stems from the fact that any attempt by the attacker to increase the distance between the node and verifier will be detected by the verifiers. This is due to the property that, when a node falls within a triangle, if the distance to one of the verifiers is enlarged, the node must appear closer to at least one of the other verifiers that form the vertices of the triangle. However, the attacker cannot decrease the distance of the node to any other verifier because of distance bounding. Hence, any increase in distance by a malicious node or a false distance advertisement by a compromised node will result in the node falling outside the triangle and hence will be detected by the three verifiers formed by a triangle. For three-dimensional positioning, four verifiers forming a triangular pyramid will be required to determine the position of any node. Capkun and Hubaux [17] also propose a system of networked sensors, called SPINE, that perform distance bounding among themselves. These distance bounds are sent to a central node that calculates the positions of all the nodes. The authors use an iterative algorithm proposed in Savvides et al. [11] for this purpose.

This technique requires that processing times be on the order of nanoseconds. A typical sensor node operates at clock cycles in the order of a few hundred MHz and hence the applicability of this scheme to existing sensor hardware needs to be justified. Also, the SPINE architecture necessitates centralized reporting of distance bounds. This will increase the communication overhead, and the centralized node will result in a single point of failure in the whole system.

12.5.2 Secure Range Independent Localization

Lazos and Poovendran [18] propose a secure localization scheme called *Serloc.* Serloc does not employ ranging to estimate node positions. The Serloc architecture consists of strategically placed nodes called *locators.* The locator nodes are equipped with directional antennas and they know their positions accurately. The locator nodes also have a range R that is much greater than the range of a sensor node. This serves to decrease the locators necessary for a particular area. The locators advertise their location in all their sectors, where a location advertisement is of the form

$$L_i : (X_i, Y_i)||(\theta_1, \theta_2)$$

where (X_i, Y_i) corresponds to the locator position and (θ_1, θ_2) corresponds to the sector of transmission. Any sensor node in the network hears from multiple locators. With the location and sector information of all the locators heard, the sensor forms a possible sample space within which it may be located. This region will be determined based on the locator positions and will be divided into a grid consisting of cells. For every grid in the cell, the node computes a vote based on its estimate of the number of locators whose signal covers the cell. Thus, a grid score table is formed that comprises the scores of all the cells. The position of the node is calculated as the centroid of all the cells with the maximum score. Figure 12.4 illustrates localization by Serloc. Two locators $L1$ and $L2$ advertise their positions along with sector ranges to the node u. In reality, many more locators will be required for localization.

Serloc provides authentication of location advertisements through a combination of cryptographic key sharing and one-way hash functions. A global symmetric key is shared between every node in the system. Nodes also store a pair-wise key for every locator in the system. The global key is used in location advertisements, whereas the pair-wise keys are used in location resolution in the presence of attacks. Serloc guards itself against wormhole attacks through the following two properties:

1. *Sector uniqueness property:* A node can be physically present in the range of only one sector of a locator at a time. Hence, when any node hears two messages corresponding to different sectors of the same locator (through the wormhole), the node detects an attack.

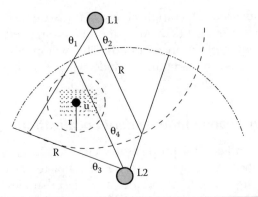

Figure 12.4   Position estimation in Serloc, where node *u* determines its position by hearing location advertisements from *L*1 and *L*2.

2. *Communication range property:* A node cannot hear from any two locators that are $2R$ apart. This is due to the bound on the communication range of locators, R. Hence, a node that detects locaters that are $2R$ apart concludes that it is under attack.

Serloc is inherently resistant to the Sybil attack [14] because sensors do not rely on other sensors to determine their position. Hence, no sensor that impersonates another sensor can contribute to an incorrect position calculation. However, when the global key used by locators is compromised, the impersonators can replay the message. To combat this, location advertisements contain a chained hash message along with the actual location advertisement. This enables detection of duplicate location advertisements by a sensor that has already heard from the original locator. Even a sensor that has not heard from the original locator can detect malicious advertisements through the wormhole detection techniques. Once an attack is detected, a collision resolution algorithm that uses the secret pair-wise keys between the node and the locator is used for location determination.

Serloc [18] provides protection against the common secure positioning attacks, namely wormhole, replay, and Sybil attacks. The localization scheme proposed also outperforms existing range-free algorithms at the cost of using directional antennas. It must be noted that this technique does not provide protection against compromised locators and, hence, attacks can be targeted at the locators in an effort to compromise them.

12.5.3   Statistical Filtering of Malicious Signals

Li et al. [9] illustrate the use of statistical filtering techniques for secure positioning. The process of locationing is typically disrupted by attackers who mutate the legitimate localization data by the addition of noise-like signals. In range-based schemes, this can be done by distance modification and attacks such as wormhole are employed in range-free schemes to accomplish the addition of unwanted signals. Hence, one way of securing the localization process is to employ robust statistical estimation techniques that can withstand the unwanted data introduced by the attackers. If the system is able to filter out the malicious signals, the localization process can then be performed only using legitimate signals, thus providing accurate locationing.

As mentioned, a popular technique for localization is triangulation. We provide a brief discussion of triangulation here to explain statistical techniques for filtering out malicious signals. A node that wishes to establish its position by triangulation typically collects a set of (x, y, d) values from beacons positioned at (x, y). Here, d is the node–beacon distance and can be estimated using range-based and range-free techniques described earlier. If (x_0, y_0) is the estimated node position, this estimate can be calculated by minimizing the mean-square error arising from the range estimates; ϵ, the error in location estimation from a single beacon, is given by

$$\epsilon = d - \sqrt{(x - x_0)^2 + (y - y_0)^2}$$

Minimizing this error term for the (x, y, d) vectors heard from all the beacons result in the position estimate (x_0, y_0).

The overall function to be minimized is

$$\delta = \sum_{j=1}^{j=k} \frac{d - \sqrt{(x_j - x_0)^2 + (y_j - y_0)^2}}{k}$$

This method works very well when the location estimates d are accurate. However, in the presence of a malicious adversary, these estimates are corrupt and hence when the error term is minimized by including these inaccurate values, the position estimate (x_0, y_0) is incorrect. The authors propose the use of Least Median Squares (LMS) [21] in lieu of the Least Squares scheme described above to effectively filter out malicious signals. If N is the overall sample space, LMS involves choosing M random subsets of size n each and then calculating the position for each of these M subsets. The final position estimate is obtained by taking the median of values returned by these M subsets. For example, (x_0, y_0) can be estimated by minimizing the error function δ as follows:

$$\delta = med_j[d - \sqrt{(x_j - x_0)^2 + (y_j - y_0)^2}]^2$$

LMS is a computationally intensive algorithm that results in a more accurate solution in the presence of malicious signals. Hence, the authors also propose an online algorithm that switches between the LMS method and LS method depending on the prevailing circumstances. The decision to switch between LS and LMS is made based upon the variance of the data that is observed. For example, in the presence of malicious signals, the variance in position estimate indicated by individual samples will be large. Thus, based on the variance level observed, the technique can choose between LMS (for high variance data) and LS (for low variance data). The authors propose this method to reduce computational complexity on the sensor nodes.

Thus, the above proposed technique secures localization by means of statistical filtering. It can also be seen that the scheme does not warrant the use of cryptographic quantities and hence reduces the infrastructure demands on the sensor network. On the other hand, this scheme is not resistant to attacks of significant strength and might lead to incorrect location estimates. Also, attacks such as Sybil that enable attackers to take on multiple identities can dramatically increase the number of attackers and hence introduce more noise into the system than what the technique can handle. This scheme can be strengthened by the addition of a complementary cryptographic infrastructure that can provide resistance against attacks such as Sybil [14]. On the whole, statistical filtering is an interesting idea and its efficacy should be examined in much more detail.

12.5.4   Range-Based Voting

Liu et al. [19] propose a localization technique that determines the node position in the presence of malicious adversaries. In this scheme, beacons present in the network help the non-beacon nodes determine their positions in the following manner. Beacon nodes advertise their location to all the non-beacon nodes in communication range by broadcasting the location advertisement. To avoid the spread of malicious location advertisements, the beacon advertisements are authenticated by means of broadcast authentication scheme similar to the one used in Perrig et al. [22]. Any node that wants to determine its position does so by listening to beacon advertisements. The node also determines the distance between itself and the beacons whose advertisement it hears through a ranging technique. The authors use a signal strength based scheme, although other schemes such as ToA and TDoA can also be used. Because ranging based on signal-strength is used, malicious nodes can easily perform distance modification and hence compromise the localization process. To combat this, the authors propose a two-dimensional grid space divided into cells of equal area. The node also maintains counters corresponding to all cells, and these counters are initialized to zero. When a node hears from a beacon, it calculates d, the distance between itself and the beacon. Based on the knowledge of d and ϵ, the average error in location estimation, the node determines a

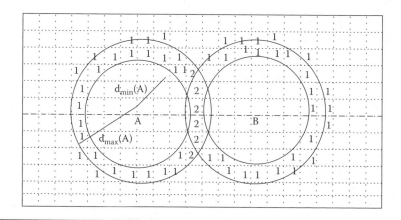

Figure 12.5 Voting-based localization.

possible range of cells in the grid that can hear the beacon. In Figure 12.5, when a node hears a beacon from A, it determines that it can fall within the area bounded by the circles of radii d_{min} and d_{max}. For all the cells that fall within this area, the counter is incremented by one. A similar process is carried out for all the beacons that the node hears from. After taking the location advertisements from all beacons into consideration, the node finally determines the cells that have the maximum value in the counter, and hence the maximum vote. The centroid of all these cells yields the correct position estimate. From Figure 12.5, it can be seen that the cells that fall within the range of both beacons A and B have a vote of 2, whereas the cells that fall within the range of a single beacon have a vote of 1. The node has a very high probability of being in cells with the vote of 2 because it heard from both beacons. It can be seen that the basic scheme is not highly accurate because the number of cells must be kept small to save memory at the sensor nodes. Accuracy can be improved using an iterative refinement scheme. Once the cells with maximum votes are determined, the area covered by these cells can be further subdivided into smaller cells, thus providing higher granularity. The voting process can now be performed on these smaller-sized cells, thus providing increased accuracy.

The security of the scheme is based on the fact that the number of malicious nodes will be less than the legitimate beacons that will dominate the voting process. In this scenario, the false location advertisements/distances introduced by the adversaries are outweighed by the legitimate beacons. Because the nodes use, at most, one location reference from a beacon, the authors claim that the number of malicious nodes must be higher than the number of legitimate beacons to disrupt the localization process. An interesting attack on this scheme is the wormhole attack, which will result in the replay of signals from distant locations. The authors advocate the use of a wormhole detection technique such as packet leashes [13] to guard against these attacks.

12.5.5 Location Verification

In many situations, validation of location claims of a user is of paramount importance. One example is location-based access control where a user gets access to certain privileges based on his location. In this case, the node that claims a particular location is called the prover and the node that verifies the location claim is called the verifier. Sastry et al. [23] illustrate a mechanism to verify the location claims of a user. In this technique, the prover sends a message to a verifier advertising its location. The verifier then sends an encrypted random nonce to the prover through a radio link. The prover then sends back a reply to this nonce through an ultrasound link. The verifier now estimates the distance between itself and the prover by measuring the round-trip propagation time. This technique is similar to distance bounding with the radio link replaced by ultrasound in one direction. The use of ultrasound here is to make the processing time at the prover comparable to the round-trip propagation time. If the processing time is not comparable to the round-trip propagation time, then the system cannot resolve the distance accurately due to uncertainty in the processing time. This system ascertains the presence of the prover in a particular area through multiple verifiers. Due to the use of the ultrasound link, the distance between the prover and the verifier can be reduced by a malicious node by speeding up the link with the use of radio frequency and hence the system mandates that there should be no adversaries in the region where the prover tries to prove its presence. On the whole, this system is far more practical than SPINE [17] due to the use of the ultrasound link, but imposes additional constraints on the attacker capabilities as a result.

12.5.6 Comparison of Existing Schemes

In this section we compare the existing localization techniques and provide an analysis of their performance with regard to various parameters such as accuracy, architecture, etc. Table 12.1 compares the performance on non statistical filtering based techniques. We do not consider statistical filtering techniques as they are generic and do not fit well into the comparison framework presented here. Also, statistical filtering techniques can be used independent of the exact localization methodology (range-based or range-free) used.

One of the most important parameters of interest is the locationing accuracy because the accuracy of the system can have a direct bearing on the applications that use the location information. We observe that both Serloc and SPINE algorithms have a high degree of accuracy, whereas the

Table 12.1 Comparison of Existing Schemes

Parameter	Serloc	SPINE	Voting-Based
Ranging	Range-free	Dependent	Dependent
Accuracy	High	High	Low
Number of beacons	Low	Low	High
Constraints	Requires directional antennas	Requires very fast processing time	None
Architecture	Distributed	Centralized	Distributed
Creation of new beacons	No	Yes	Possible
Compromised beacons	Not resistant	Not resistant	Resistant
Computational complexity	Not a factor (beacons are powerful)	High	Low
Distance modification attack	Not possible	Not possible	Resistant
Wormhole attack	Resistant	Resistant	Not resistant
Sybil attack	Resistant	Resistant	Not resistant

voting-based technique has a lower degree of accuracy. Serloc's accuracy is boosted by its use of directional antennas, and SPINE is more accurate because of the precise ranging hardware employed in position estimation. In the case of the voting-based scheme, although accuracy is not high, locationing performance degrades gracefully in the face of increased attacker strengths. It is interesting to consider the number of beacons necessary to determine the node position. Both the Serloc and SPINE algorithms require a minimum of three beacons for localization. As the number of beacons increases, the accuracy increases. The voting-based scheme requires a significantly higher number of beacons because of the fact that it requires the number of beacons to be higher than the number of malicious nodes. This, coupled with the fact that the algorithm uses approximation in the form of voting, necessitates the need for an increased number of beacons. It must also be noted that the increased accuracy and the reduced number of beacons in both Serloc and SPINE are achieved with the help of additional resources that are part of the system. In the case of Serloc, the directional antennas help in improving the accuracy and help in resisting certain types of attacks. SPINE, on the other hand, requires processing power to be bounded to the order of nanoseconds. This makes SPINE a computationally intensive technique, and it remains to be seen whether sensor nodes can really achieve processing times on the order of nanoseconds.

The architecture of the localization system is an important design decision and has the potential to significantly alter the efficacy of the system. As mentioned, distributed operation modes are preferred in sensor nodes due to near-uniform traffic distribution over the entire network. From a security standpoint, distributed systems augur well due to the absence of a single point of failure, which is a characteristic of centralized systems. It can be seen that Serloc and the voting-based schemes compute node positions in a completely distributed fashion, whereas SPINE uses a centralized authority to compute node positions. In distributed systems, an important classification is whether the nodes are homogeneous (of similar capabilities) or heterogeneous (of different capabilities). In comparing these schemes, with respect to homogeneity, it can be seen that all the three schemes require beacons that already know their node positions. At any point in time, the network has two kinds of nodes: the beacons and the ordinary nodes. It will be interesting to analyze the learnability of the techniques used for localization. Whenever a node determines its own position, it can theoretically act as a beacon because the knowledge of location alone differentiates the beacon from the ordinary node. If a system can use this property of generation of new beacons when existing nodes determine their position, the system is said to be "learnable." SPINE uses an iterative multilateration algorithm that creates new beacons out of ordinary nodes. In the voting-based scheme too, a node can become a beacon once it finds out its own position. This property cannot be exploited in Serloc because the beacons in Serloc are powerful nodes with directional antennas. Ordinary sensor nodes do not satisfy this criteria. Hence, Serloc cannot create new beacons in the network as nodes discover their own positions.

In the context of sensor networks, node compromise is always a possibility as it is expensive to protect the nodes against compromise. Hence, it would be better if the localization scheme is immune to node compromise. Both SPINE and Serloc algorithms are not resistant to compromised beacons. Hence, when beacons are compromised, the performance of these techniques degrades significantly. On the other hand, the voting-based scheme is resistant to compromised beacons because a majority vote is taken to determine node positions. In Section 12.3, we enumerated the different attacks on wireless sensor networks, and it would be interesting to see the performance of the above schemes under the attacks. All three schemes are resistant to distance modification attacks. In fact, distance modification attacks are not possible against Serloc as it is range-free. SPINE handles a distance modification attack by the point in triangle test and the voting-based scheme is resistant to distance modification through the use of the majority voting scheme. In the case of a wormhole attack, the voting-based scheme will get false location advertisements routed through the wormhole. The scheme is not resistant to wormhole attack and will result in incorrect position estimates in the absence of wormhole detection techniques such as packet leashes [13]. As mentioned, attackers employing the Sybil attack typically take different node identities at different points in time. Hence, a single malicious attacker can take different identities and publish false location advertisements. This will lead to an attack of significant strength against the voting-based scheme as it is based on a majority voting scheme and the number of attackers is multiplied manifold in a Sybil attack. The other two schemes are resistant against Sybil attacks through the use of adequate cryptographic entities.

12.6 Conclusions

This chapter described the problem of secure localization and presented a survey and comparison of some of the existing schemes in literature. The strengths and weaknesses of each such scheme were discussed. This topic of research is in its nascent stages, and we hope to see more active research on this topic in an effort to gain a better understanding of the secure localization problem and the solutions.

References

1. Y.-B. Ko and N.H. Vaidya, Location-aided routing (LAR) in mobile ad hoc networks, in *Proceedings of the 4th Annual ACM/IEEE International Conference on Mobile Computing and Networking (MobiCom)*, Dallas, TX, pp. 66–75, 1998.
2. S. Basagni, I. Chlamtac, V.R. Syrotiuk, and B.A. Woodward, A distance routing effect algorithm for mobility (DREAM), in *Proceedings of the 4th Annual ACM/IEEE International Conference on Mobile Computing and Networking (MobiCom)*, Dallas, TX, pp. 76–84, 1998.
3. P. Bahl and V.N. Padmanabhan, RADAR: An in-building RF-based user location and tracking system, in *Proc. IEEE INFOCOM*, Tel Aviv, Israel, pp. 775–784, 2000.
4. N.B. Priyantha, A. Chakraborty, and H. Balakrishnan, The cricket location-support system, in *Proceedings of the 6th Annual International Conference on Mobile Computing and Networking (MobiCom)*, Boston, MA, pp. 32–43, 2000.
5. R. Want, A. Hopper, V. Falcao, and J. Gibbons, The active badge location system, *ACM Transactions on Information Systems*, Vol. 10, pp. 91–102, January 1992.
6. N. Bulusu, J. Heidemann, and D. Estrin, GPS-less low cost outdoor localization for very small devices, *IEEE Personal Communications Magazine*, Vol. 7, pp. 28–34, October 2000.
7. T. He, C. Huang, B.M. Blum, J.A. Stankovic, and T. Abdelzaher, Range-free localization schemes for large scale sensor networks, in *Proceedings of the 9th Annual International Conference on Mobile Computing and Networking (MobiCom)*, San Diego, CA, pp. 81–95, ACM Press, 2003.
8. D. Niculescu and B. Nath, DV based positioning in ad hoc networks, *Telecommunication Systems*, Vol. 22, No. 1–4, pp. 267–280, 2003.
9. Z. Li, W. Trappe, Y. Zhang, and B. Nath, Robust statistical methods for securing wireless localization in sensor networks, in *Proceedings of the Fourth*

International Symposium on Information Processing in Sensor Networks (IPSN), Los Angeles, CA, 2005.

10. M. Youssef and A. Agrawala, The Horus WLAN location determination system, in *ACM International Conference on Mobile Systems, Applications, and Services (MobiSys)*, Seattle, WA, 2005.

11. A. Savvides, C.-C. Han, and M.B. Srivastava, Dynamic fine-grained localization in ad-hoc networks of sensors, in *Proceedings of the 7th Annual International Conference on Mobile Computing and Networking (MobiCom)*, Rome, Italy, pp. 166–179, 2001.

12. E. Elnahrawy, X. Li, and R. Martin, The limits of localization using signal strength: a comparative study, in *IEEE SECON*, Santa Clara, CA, 2004.

13. Y.-C. Hu, A. Perrig, and D.B. Johnson, Packet leashes: a defense against wormhole attacks in wireless networks, in *Proc. IEEE INFOCOM*, San Francisco, CA, 2003.

14. J. Douceur, The sybil attack, in *IPTPS02 Workshop*, Cambridge, MA, 2002.

15. J. Newsome, E. Shi, D. Song, and A. Perrig, The Sybil attack in sensor networks: analysis & defenses, in *Proceedings of the Third International Symposium on Information Processing in Sensor Networks (IPSN)*, Berkeley, CA, pp. 259–268, 2004.

16. I. Akyildiz, W. Su, Y. Sankarasubramaniam, and E. Cayirci, A survey on sensor networks, *IEEE Communications Magazine*, Vol. 40, No. 8, pp. 102–114, 2002.

17. S. Capkun and J.-P. Hubaux, Secure positioning of wireless devices with application to sensor networks, in *Proc. IEEE INFOCOM*, Miami, FL, March 2005.

18. L. Lazos and R. Poovendran, Serloc: secure range-independent localization for wireless sensor networks, in *Proceedings of the 2004 ACM Workshop on Wireless Security (WiSe)*, Philadelphia, PA, pp. 21–30, 2004.

19. D. Liu, P. Ning, and W.K. Du, Attack-resistant location estimation in sensor networks, in *Proceedings of the Fourth International Symposium on Information Processing in Sensor Networks (IPSN)*, Los Angeles, CA, 2005.

20. S. Brands and D. Chaum, Distance-bounding protocols, in *EUROCRYPT '93: Workshop on the Theory and Application of Cryptographic Techniques on Advances in Cryptology*, Lofthus, Norway, pp. 344–359, 1994.

21. P. Rousseeuw and A. Leroy, *Robust Regression and Outlier Detection*, Wiley, 2003.

22. A. Perrig, R. Szewczyk, V. Wen, D. Culler, and J.D. Tygar, SPINS: security protocols for sensor networks, *Wireless Networks*, Vol. 8, pp. 521–534, Sept. 2002.

23. N. Sastry, U. Shankar, and D. Wagner, Secure verification of location claims, in *ACM Workshop on Wireless Security (WiSe)*, San Diego, CA, September 2003.

Chapter 13

Cross-Layer Design for the Security of Wireless Sensor Networks

Mingbo Xiao, Xudong Wang, and Guangsong Yang

Contents

13.1   Abstract

The wireless sensor network (WSN) is a newly emerging technology that represents a significant improvement over traditional sensors in many applications. Improved security is especially important for the success of the WSN, because the data collected is often sensitive and the network is particularly vulnerable. A number of approaches have been proposed to provide security solutions against various threats to the WSN, most of which are based on the layered design. In this chapter we point out that these layered approaches are often inadequate and inefficient in addressing the distinguishing features of the WSN, such as open access medium, dynamic network topology, and limited node resources on computation, storage, bandwidth, and power. Instead, it is advantageous to break with the conventional layering rules and design a security scheme for the WSN based on information from several protocol layers. We overview the existing schemes for the cross-layer design of WSN security and propose some new solutions. A few open problems in this area are also discussed.

13.2   Introduction

Recent advances in wireless communications and electronics have enabled the development of wireless sensor networks (WSN), which comprise many low-cost, low-power, and multifunctional sensor nodes to accomplish certain sensing tasks in an intelligent manner. These sensor nodes are small in size and consist of sensing, data processing, and communicating components, and hence represent a significant improvement over traditional sensors [1,2]. Owing to the above-described features of sensor networks, a wide range of applications has been recognized in areas such as ocean and wildlife monitoring, manufacturing machinery performance monitoring, building safety and earthquake monitoring, and military applications.

While different aspects of sensor networks have been under intense research, most efforts have focused on energy efficiency, network protocols, and distributed databases [3,4], and relatively few results have been reported on the security of sensor networks, which are also very important in many applications, especially in battlefield applications, premise security and surveillance, and some critical systems such as airports, hospitals, etc.

In general, a network cannot work well or becomes useless if there is no sufficient security mechanism to protect the privacy and integrity of the data. Although different applications may require different security levels, there are four fundamental security requirements, namely:

1. *Availability*: The service offered by the node will be available to its users when expected.
2. *Authenticity of origin*: The principals with whom one interacts are the expected ones.
3. *Authentication of data (integrity)*: The data received should be authentic and not tampered.
4. *Confidentiality (privacy)*: The information exchanged should be understood by the expected users alone. This is often realized by encrypting the messages with a key that is usually made available by the authentication process.

For a network like WSN, two additional requirements are

1. *Survivability*: The ability to provide a minimum level of service in the presence of power loss, failures, or attacks.
2. *Degradation of security services*: The ability to change security level as resource availability changes.

It is more difficult to provide an efficient and scalable security solution for sensor networks because of their distinguished characteristics [5,6]:

1. Vulnerability of channels owing to shared wireless medium
2. Vulnerability of nodes in an open network architecture
3. Absence of infrastructure
4. Dynamic change in network topology
5. Hostile deployment environments
6. Resource limitations (computation capacity, memory, power, etc.) on sensor nodes
7. Very large number and dense distribution of nodes

The vulnerability and constraints of sensor networks offer the attacker unique attacks that are not found in traditional networks. Various kinds of active and passive attacks have been recognized:

1. *A denial-of-service attack* for the purpose of battery power exhaustion. For example, a malicious node could prohibit another node from going back to sleep, thus causing the battery to be drained [7].
2. *Eavesdropping and invasion* become fairly easy for wireless communication if no proper security measures are taken. An adversary

could easily extract useful information from conversations between nodes. With the information, a malicious user could join the network undetected by impersonating as some other trusted node, to have access to private data, disrupt the normal network operations, or trace the actions of any node in the network.

3. *Physical node tampering* leading to node compromising.
4. *Forced battery exhaustion* on a node.
5. *Radio jamming* at the physical layer.

Security techniques with considerations for the characteristics of sensor networks should be devised in light of these attacks. Because vast quantities of sensor nodes are distributed in the sensor network, extremely low cost and low power become the core design challenges. The low cost constrains the resources that can be implemented on the devices, and the low power requires the operations to be done in a highly efficient way. Moreover, owing to the large-scale and distributed nature of WSNs, the protocols and algorithms must be scalable. Quite recently, a number of solutions have been proposed specifically for securing WSNs [8–10]. Most of the solutions deal with attacks on one protocol layer, but we will argue that the so-called layered scheme is inadequate in providing security for sensor networks; instead, cross-layer solutions are needed to improve the performance.

13.3   Layered Security Schemes

Sensor networks are usually deployed in unattended or even hostile environments (such as battlefields). Wireless channels are used to communicate messages or exchange routing information. Sensor nodes are susceptible to physical capture. Moreover, there are various constraints in computation, memory, and power resources in the nodes. Hence, sensor networks are vulnerable to various kinds of attacks on one or many layers of the communication protocols, and security is a challenging task for sensor networks.

The problem of securing ad hoc networks has received a great deal of attention in the literature [11–13]. Unfortunately, WSNs are sufficiently different in their characteristics from ad hoc networks, so some security solutions designed for the former may not apply automatically to the latter. The main differences lie in several aspects [1,14]. The number of sensor nodes in a sensor network can be several orders of magnitude greater than the nodes in an ad hoc network. Sensor nodes are densely deployed and are prone to failure. These nodes mainly use a broadcast communication paradigm, whereas most ad hoc networks are based on point-to-point communications. Sensor nodes are even more limited in resource such as power, computational capacity, and memory. When designing security protocols for a WSN, we should place emphasis on power efficiency, node density and availability, adaptability, self-configurability, and simplicity.

13.3.1 Networking Protocols

Similar to MANET, the operation of a sensor network is mainly enabled through ad hoc routing, MAC protocols, and certain physical layer techniques. Therefore, attacks on the functions in these protocol layers appear more frequently than attacks in other protocol layers.

13.3.1.1 Routing Protocols

Most existing routing protocols proposed for sensor networks are designed not for security, but for optimizing the limited resources. Moreover, sensor networks cannot provide resources available to traditional networks for security. The adversary can effectively mount miscellaneous attacks at the networking layers, to prevent an intended recipient from receiving the routing messages. The attackers may advertise routing updates that mismatch the requirements of a routing protocol, as explained in Hu et al. [15] and Zapata and Asokan [16]. Another type of attack is on the function of packet forwarding or routing protocols [17] or perform selective forwarding or packet dropping [18]. The attacker may not change routing tables, but lead the packets on the routing path to a different destination that is not consistent with the routing protocol. Moreover, the attacker may sneak into the network and then impersonate a legitimate node and does not follow the required specifications of a routing protocol [19]. Some malicious nodes may create wormhole and short-cut the normal flows among legitimate nodes [20]. Multi-path routing is a possible defense against this type of attack [21]. The basic idea is to use multiple disjoint paths to route a message such that it is unlikely that every path is controlled by compromised nodes.

13.3.1.2 MAC Protocols

Similar attacks as in routing protocols can also occur in MAC protocols. For example, the backoff procedures and NAV for virtual carrier sense of IEEE 802.11 MAC may be misused by some attacking nodes so that the network is always congested by these malicious nodes [22]. Attackers can sneak into the network by misusing the cryptographic primitives [23].

Two approaches have been considered to improve security at the MAC layer. One is the secure MAC protocol, in which the whole MAC mechanism has already taken into account security measures. The other is to implement a security monitoring tool that can interactively work together with MAC to detect the attacks and prompt responses. Because the monitoring tool is MAC specific, a tool developed now is usually not able to explore the possible issues in future MAC protocols. However, certain features can be considered in the architecture so that the monitoring tool could be extended to adapt to new MAC protocols.

13.3.1.3 Physical Layer Techniques

Because communications in sensor networks are on wireless channels, the physical layer is the most vulnerable part that is prone to attack. Attack on the physical layer, such as intentional interference or jamming, can easily force a node or even the whole network out of service. It should be noted that without a secure physical layer, security efforts in higher protocol layers may turn out to be in vain, because all higher layers depend on the physical layer to deliver information. Thus, certain techniques in the physical layer must be taken into account if a highly secure sensor network is demanded. These techniques include ultra-wide band (UWB), spread spectrum, and smart antennas.

13.3.2 Security Issues in Key Management

The problem of distributing and updating the cryptographic keys to valid members is known as *key management*. Key management is one of the most important tasks in cryptographic mechanisms for networks. However, for sensor networks, key management is much more challenging because there may be no central authority, trusted third party, or server to manage security keys. In this case, key management must be performed in a distributed way.

A self-organization scheme to distribute and manage the security keys is proposed in [4]. In this system, certificates are stored and distributed by the users themselves. When the public keys of two users must be verified, they first merge the local certificate repositories and then find the appropriate certificate chains within the merged repositories that can pass this verification. This self-organization key management scheme was based on an impractical assumption: that is, trust is transitive. Thus, more work is needed to enhance this scheme.

Use of a single shared key in a whole WSN is not a good idea because an adversary can easily obtain the key. Thus, sensor nodes must adapt their environments, and establish a secure network by (1) using pre-distributed keys or keying materials, (2) exchanging information with their immediate neighbors, or (3) exchanging information with computationally robust nodes.

Although there is ongoing work [25–27] to customize public key cryptography and elliptic key cryptography for low-power devices, such approaches are still considered costly, owing to the high processing requirements. Key distribution and management problems in WSN are difficult ones and require new approaches.

13.3.3   Security Issues in Cryptographic Protocols

In a cryptographic protocol, exchange of information among users happens frequently. To avoid being caught in a disadvantageous situation, exchanging parties usually employ a fair exchange protocol that depends on a trusted third party. However, this trusted party is not available in sensor networks owing to lack of infrastructure. Thus, another exchange scheme, called rational exchange, must be used. Rational exchange ensures that a misbehaving party cannot gain anything from misbehavior, and thus will not have any incentives to misbehave [28]. However, rational exchange cannot really eliminate an enemy's incentives of misbehavior, as the enemy may just focus on forcing the network not being able to work.

13.3.4   Service Availability

In a sensor network, owing to the lack of central coordination and network dynamics, cooperation and collaboration are key to guaranteeing services uninterrupted. Thus, misbehavior of a node, either intentionally or unintentionally, may cause service degradation or even shutdown. In some cases, the reason for the existence of misbehavior is that network protocols such as MAC and routing protocols did not or were not able to be taken into account security in the design phase. Thus, protocol enhancement and additional security measures are usually needed to improve security. In other cases, security solutions for sensor networks may contain some security holes, which leave a door for malicious users to explore the possibility of compromising the security of mobile ad hoc networks. This is especially common in battlefield applications.

13.4   Limitations of Layered Security Approaches

Because sensor networks pose unique challenges, traditional security techniques used in traditional networks cannot be applied directly. First, to make sensor networks economically viable, sensor devices are limited in their energy, computation, and communication capabilities. Second, unlike traditional networks, sensor nodes are often deployed in accessible areas, thus presenting the added risk of physical attack. Finally, sensor networks interact closely with their physical environments and with people, thereby posing new security problems. Consequently, existing security mechanisms are inadequate, and new ideas are needed.

Owing to resource limitations on computation, storage, and bandwidth, the following aspects should be carefully considered when designing a security scheme:

1. *Power efficiency*: Energy supply is scarce and hence energy consumption is a primary metric to consider.

2. *Node density and reliability*: WSNs have to scale to much larger numbers (thousands, hundreds of thousands) of entities than current ad hoc networks, requiring different, more scalable solutions. Sensor nodes are prone to failures. Unfortunately, existing security designs can address only a small, fixed threshold number of compromised nodes; the security protection completely breaks down when the threshold is exceeded [29].

3. *Adaptive security*: With numerous combinations of sensing, computing and communication technologies, WSNs are conceivable with very different network densities — from very sparse to very dense deployments. They must interact with the environment, and the traffic characteristics can be expected to be very different from other, human-driven forms. Thus they will require different or at least adaptive security protocols.

4. *Self-configurability*: Also similar to ad hoc networks, WSNs will most likely be required to self-configure into connected networks. However, the difference in factors such as traffic and energy trade-offs may require new solutions. For example, sensor nodes may have to learn about their geographical position [30].

5. *Simplicity*: Because sensor nodes are small and energy is scarce, the operating and networking software must be kept orders of magnitude simpler as compared to today's desktop computers.

6. *May not have global ID like IP address*: This is owing to the fact that the global ID will cause a large amount of overhead due to a large number of sensors.

To effectively address the above issues, it may be advantageous to break with conventional layering rules for networking software. For illustration, next we first analyze some important limitations of the layered security approaches.

13.4.1 Redundant Security Provisioning

A WSN system is subject to a large number of attacks, and each security mechanism consumes some precious WSN resources (battery, memory, computation power, and bandwidth). The provision of maximum security services in each sensor node [9] can lead to unnecessary waste of system resources, and can significantly reduce the network lifetime. It is observed that without a systematic view, individual security protocols developed for different individual protocol layers might provide redundant security services, and hence consume more WSN resources than necessary [31]. In some sense, an unorganized design of security provisioning may use up network resources and therefore unintentionally launch a DoS attack, which is referred to as a security service DoS (SSDoS) attack.

In general, there may be several protocol layers within the network protocol stack capable of providing security services to the same attack. In this case, when the original data goes through the protocol stack from the top-most layer, it will be processed layer-by-layer. Some part of the data packets may go through the security-provision operations of different layers and result in redundant security provisioning.

13.4.2 Non-adaptive Security Services

Because attacks on a WSN come from any layers in any protocol, a counter-attack scheme in some protocol layer is unlikely to guarantee security all the time. For example, link layer security typically addresses confidentiality provisioning, two-party authentication, and data freshness, but no security problems in the physical layer. However, an insecure physical layer may practically render the entire network insecure. It is easy to understand that multi-layer solutions or cross-layer solutions can achieve better performance. Furthermore, additional security capability can be achieved via self-adaptive security services because they are flexible in dealing with the dynamic network topology as well as various types of attacks.

13.4.3 Power Inefficiency

In designing a sensor network, a very important problem to consider is *energy efficiency*. There are several sources of power consumption in sensor networks, including idle listening, retransmissions resulting from collisions, control packet overhead, and unnecessarily high transmitting power. Correspondingly, there are different methods of reducing power consumption. Some approaches limit the transmission power, aiming to increase the spatial reuse while maintaining network connectivity [32]. At the network layer, power-aware routing protocols demonstrate significant power savings [33]. There are approaches at the MAC layer [34,35] that turn the wireless transceivers off whenever possible, to reduce the idle listening power as well as the number of collisions. Depending on the specific application, several approaches at the application layer [36,37] can dramatically improve power consumption. To reduce the power consumption of every node, a method was proposed in Yu and Guan [38] to drastically reduce the number of potential neighbors of each node. It becomes increasingly clear that power efficiency design cannot be addressed completely at any single layer in the networking stack [39].

13.5   Cross-Layer Security Solutions

A set of four guiding principles have been proposed in Jones et al. [40] to address the problem of securing WSNs.

1. The security of a network is determined by the security over all layers.
2. In a massively distributed network, security measures should be amenable to dynamic reconfiguration and decentralized management.
3. In a given network, at any given time, the cost incurred due to the security measures should not exceed the cost assessed due to the security risks at that time.
4. If the physical security of nodes in a network is not guaranteed, the security measures must be resilient to physical tampering with nodes in the field of operation.

The first point is actually on cross-layer security design. Owing to its extreme vulnerability, satisfactory security provisioning in a WSN is crucial. However, as discussed in previous sections, security based on layered design is often inadequate. Moreover, a highly secure mechanism inevitably often consumes a rather large amount of system resources, which in turn may unintentionally cause a security service denial-of-service attack. As a result, the cross-layer design is believed to provide a better security solution. For example, energy is a physical layer parameter, while security is an application layer service, so the energy-efficient security design must take the cross-layer interaction into consideration.

13.5.1   Existing Cross-Layer Design Schemes

There are a few existing schemes of cross-layer security design that consider factors from different protocol stacks. We will first briefly review some categories of typical schemes in this section.

The first category is for *key management*. A cross-layer design approach was introduced in Lazos and Poovendran [41] for key management in wireless multicast. With this approach, cryptographic keys to valid group members are distributed in an energy-efficient way. By considering the physical and network layers in combination, an optimization problem was formulated in Eschenauer and Gligor [47] to minimize the energy required for re-keying. The authors then present a sub-optimal, cross-layer algorithm that considers the node transmission power (physical layer property) and the multicast routing tree (network layer property) in order to construct an energy-efficient key distribution scheme (application layer property). Another cross-layer design algorithm was presented in Lazos et al. [42] for multicast key distribution that uses routing energies from the sender and

Hamming codes representing the paths from the sender to each node to minimize the average energy for key updates.

A solution was proposed in Jones et al. [43], using parameterized frequency hopping and cryptographic keys in a unified framework, to provide differential security services for the WSN. The solution supports a differential security service that can be dynamically configured to accommodate changing the application and network system state. The paradigm for securing sensor networks is based on a holistic approach to securing multiple layers in the protocol stack. An important aspect of this paradigm is the exploitation of the interplay between security measures in different layers to provide a security service for the entire network.

13.5.2 New Schemes of Cross-Layer Design

For security provisioning in sensor networks, each protocol layer emphasizes different aspects. The physical layer improves information confidentiality using encoding. The link layer and network layer are concerned with the encryption of the data frame and routing information. The application layer focuses on key management and exchange, which provides security support for encryption and decryption of the lower layers. When considering the security problem of sensor networks, we should take the characteristics of each layer into account, and use a cross-layer design to trade security off network performance and reduce as much redundancy as possible. For example, if the objective is to provide energy-efficient security provisioning, we might integrate the following measures: (1) at the physical layer, transmission power can be automatically tuned according to the interference strength, which reduces energy consumption and avoids congestion attacks; (2) at the MAC layer, we can limit the number of retransmissions, which prevents exhaustion attack and saves energy as well; (3) at the network layer, we can adopt multi-path routing, which avoids routing black-hole and reduces the energy consumption due to congestion.

As we discussed, the security of WSNs involves all protocol layers. Moreover, at each protocol layer, multiple functional blocks are cross-related to a security solution. Thus, it is impossible to have one cross-layer security solution that works for all protocol layers or tackles all security issues from different perspectives. One effective approach is to develop individual cross-layer security schemes for the different categories of security issues.

13.5.2.1 Cross-Layer Design for Heterogeneous Requirements and Service Types

A sensor network may include different types of sensors and perform multiple concurrent applications. Different application scenarios will have diverse security requirements. Even within an application, each individual task may have different security concerns. Slijepcevic and Potkonjak [44] classify the types of data existing in sensor networks, identify possible communication security threats according to that classification, and propose a communication security scheme in which a corresponding security mechanism is defined for each type of data. By allowing each mechanism to have different resource requirements, the authors show that this multi-tiered security architecture leads to efficient resource management, which is essential for wireless sensor networks. A link layer security framework, called SecureSense, was proposed in Qi and Ganz [45] to provide energy-efficient secure communications in sensor networks. Using runtime security service composition, SecureSense enables a sensor node to optimally allocate its resource to appropriate security services, depending on observed external environments, internal constraints, and application requirements.

These schemes have considered different security issues for different requirements and services. However, they have not taken into account the fact that differences of these services or requirements may also be reflected at different protocol layers. Different service type requires messages to be encrypted differently, and different encryption scheme also consumes different amount of energy. In addition, security overhead and energy consumption should correspond to the sensitivity of the encrypted information. Thus, we can allocate the security requirements to different layers, so as to minimize the security-related energy consumption.

13.5.2.2 Cross-Layer Design for Intrusion Detection

Research into detecting protocol intrusion has focused on routing and MAC protocols. Some research has been done on cryptographic protocols. The existing secure protocols or intrusion detection schemes are normally proposed for one protocol layer. The effect of these schemes is limited to attacks at a particular layer. They are rarely effective for attacks from different protocols layers. However, security concerns may arise in all protocol layers. Simply proposing a security scheme for one layer does not really solve the problem. Thus, it is necessary to have a security monitoring tool that contains a cross-layer detection framework that consolidates various schemes in different protocol layers. None of the existing protocols has really considered a cross-layer architecture for intrusion detection, although a preliminary framework was proposed in Zhang and Lee [46].

We also need to note that some existing solutions for one protocol layer are not well-done yet either. For example, the general assumption that MAC is for one-hop connectivity [1], which may actually not be true in sensor networks. Security schemes based on such assumptions may turn out to be obsolete in future sensor networks. Intrusion in the physical layer has always been ignored by researchers. However, this type of attack is much more serious and is difficult to detect. If a channel is intentionally jammed by malicious users, security detection schemes based on MAC or routing protocols may make it impossible to find the problem. Therefore, many research issues still remain for intrusion detection at different protocol layers.

13.5.2.3 Cross-Layer Design for Power Efficiency

As stated previously, energy conservation is an important objective of sensor network design. It is desirable to consider the energy consumption at each design stage and across protocol layers, to achieve the trade-off between energy consumption, network performance, and complexity, and maximize the lifetime of the entire network. A cross-layer approach can conserve energy while providing network security provisioning. We will demonstrate this point from the MAC layer and the network layer.

The carrier detection is liable to DoS attacks. A malicious node can take advantage of the interactions in the MAC layer to repeatedly request a channel. This not only prevents other nodes from connecting with the target node, but also can exhaust its battery due to frequent responses. From the information collected from other layers, the malicious node can be recognized, and then be isolated or limited.

At the network layer, we can choose a suitable route using information from other layers. For example, from the information of battery usage, we can choose a node with more energy left to assume more computational load for security or to relay more traffic. From authentication information, we can choose a route to steer clear of a malicious node or an attacked area. The geographical location information can help to resist attacks such as Sinkhole.

The safest and most energy-conserving node is the inactive node, that is, the node is sleeping. Various node-sleeping schemes should be exploited as much as possible.

13.5.2.4 Cross-Layer Design for Key Management

Due to the limited capacity of the sensor node, we should consider aspects that save storage space, decrease computational complexity, and reduce communication overhead for key management.

There are different key management schemes available: for example, Basic Random Key Scheme [47] and Polynomial Poll Based Key Scheme [48]. They are different in complexity, scalability, and effectiveness in resisting cracking.

Adaptive key management schemes must be devised to take into account information such as security level, congestion, location, and remaining energy. To this end, one important task is to derive the overall optimization subject to constraints across multiple protocol layers. The key management scheme based on such an optimization algorithm in turn needs to have different components located at multiple layers to work interactively to really deliver overall optimized performance.

13.5.2.5 Cross-Layer Design for Detecting Selfish Nodes

In a WSN, the connectivity of the network critically relies on the cooperation among various nodes. If one node intentionally stops forwarding packets for its neighboring nodes, the network will eventually go out of service. Such a node is called a *selfish node*. To avoid this common issue, two types of solutions are needed. One is to implement a mechanism in the communication protocols to ensure that a node has enough interest to forward packets for other nodes. The other is to develop a scheme for the communication protocols to detect selfish nodes, warn or penalize them when detected, and quickly lead them back to the collaboration mode. Both solutions heavily depend on the cross-layer design methodology, because selfish behavior can come out at any protocol layer, in particular the MAC and routing protocols. When cross-layer design is taken into account, a solution cannot only be more effective to avoid selfish behavior of one particular protocol layer, but it also tackles such problems in multiple protocol layers.

As an example for the MAC/networking cross-layer design, one component of the security scheme can be located in the network layer of a node and takes charge of monitoring packet forwarding by this node's successors. Another component of the same security scheme can be located in the MAC layer and is responsible for appending two-hop information such as two-hop acknowledgments to the standard MAC packets, and for forwarding them. Such two-hop information will be used by the upper layer component to detect selfish nodes. When it is detected, certain actions can be taken by the component in the MAC layer. Such a scheme can detect a selfish node more quickly, due to the faster actions of a MAC protocol than a networking protocol. This cross-layer architecture also reduces the communication overhead compared with a standard one-layer approach, and gives more robustness to selfish behaviors.

13.6 Conclusions

WSNs are expected to be employed in the near future in a wide variety of applications, ranging from military, to industrial, social, and domestic. They are believed to establish ubiquitous networks that will pervade society, and redefine the way in which we live and work. Security is crucial to the success of a sensor network, because people want to guarantee a high level of service availability as well as information confidentiality and integrity in the face of potential security attacks. However, sensor networks pose unique challenges in security provisioning, and traditional security techniques used in traditional networks cannot be applied directly. This is mainly due to the inherent limitations of sensor network resources such as computation capacity, power supply, and memory. The node density of sensor networks further increases the difficulty. Layered security schemes have been shown to be inadequate or inefficient, and the cross-layer designs are expected to be the solution of choice to the security of WSNs. However, many problems need further research. One is how to trade off the security level and system performance with minimal power computation. Another is to propose cross-layer interactions for detecting attacks and to provide intrusion tolerance and graceful degradation designs for network survivability. Finally, finding ways to tolerate the lack of physical layer security, perhaps through redundancy or knowledge about the physical environment, will remain a continuing overall challenge.

References

1. I.F. Akyildiz, W. Su, Y. Sankarasubramaniam, and E. Cayirci, A survey on sensor networks, *IEEE Communications Magazine*, Vol. 40 No. 8 (2002) pp. 102–114.
2. G.J. Pottie and W.J. Kaiser, Embedding the Internet: wireless integrated network sensors, *Communications of the ACM*, Vol. 43 No. 5 (2000) pp. 51–58.
3. J. Rabaey, J. Ammer, J.L. da Silva, and D. Patel, PicoRadio: adhoc wireless networking of ubiquitous low-energy sensor/monitor nodes, in *Proc. of the Workshop on VLSI* (2000) pp. 9–14.
4. D. Estrin, R. Govindan, and J. Heidemann, Embedding the Internet: introduction, *Communications of the ACM*, Vol. 43 No. 5 (2000) pp. 38–41.
5. H. Yang, H. Luo, F. Ye, S. Lu, and L. Zhang, Security in mobile ad hoc networks: challenges and solutions, *IEEE Wireless Communications*, Vol. 11 No. 1 (2004) pp. 38–47.
6. L. Buttyan and J.P. Hubaux, Report on a working session on security in wireless ad hoc networks, *ACM Mobile Computing and Communications Review*, Vol. 7 No. 1 (2002) pp. 74–94.
7. F. Stajano and R. Anderson, The resurrecting duckling: security issues in ad hoc wireless networks, in *Proc. of the 7th International Workshop on Security Protocols*, Vol. 1796 (1999) pp. 172–194.
8. A. Perrig, R. Szewczyk, V. Wen, D. Culler, and J.D. Tygar, SPINS: security protocols for sensor networks, *ACM/Kluwer Wireless Networks*, Vol. 8 No. 5 (2002) pp. 521–534.
9. TinySec, http://www.cs.berkeley.edu/ nks/tinysec/
10. A.D. Wood and J.A. Stankovic, Denial of service in sensor networks, *IEEE Computer*, Vol. 35 No. 10 (2002) pp. 54–62.
11. L. Zhou and Z.J. Haas, Securing ad hoc networks, *IEEE Network*, Vol. 13 No. 6 (1999) pp. 24–30.

12. J.P. Hubaux, L. Buttyan, and S. Capkun, The quest for security in mobile ad hoc networks, in *Proc. of ACM MOBIHOC*, Long Beach, CA, (2001) pp. 146–155.

13. J. Kong et al., Providing robust and ubiquitous security support for mobile ad hoc networks, in *Proc. of the 9th International Conference on Network Protocols, IEEE CS Press*, Los Alamitos, CA, (2001) pp. 251–260.

14. H. Karl and A. Willig, A short survey of wireless sensor networks, TKN Technical Report Series, TKN-03–018, TU Berlin (2003).

15. Y. Hu, A. Perrig, and D. Johnson, Ariadne: A secure on-demand routing protocol for ad hoc networks, in *Proc. of the Eighth Annual International Conference on Mobile Computing and Networking (MobiCom 2002)*, (2002) pp. 12–23.

16. M. Zapata and N. Asokan, Securing ad hoc routing protocols, in *Proc. of ACM WiSe*, (2002) pp. 1–10.

17. P. Papadimitratos and Z.J. Haas, Secure routing for mobile ad hoc networks, in *Proc. of SCS Communication Networks and Distributed Systems Modeling and Simulation Conference (CNDS 2002)*, San Antonio, TX, (2002) pp. 27–31.

18. C. Karlof and D. Wagner, Secure routing in wireless sensor networks: attacks and countermeasures, in *Proc. of the 1st IEEE International Workshop on Sensor Network Protocols and Applications*, Vol. 1 No. 2 (2003) pp. 293–315.

19. B. Dahill et al., A secure protocol for ad hoc Networks, in *Proc. of IEEE ICNP*, (2002) pp. 78–87.

20. Y. Hu, A. Perrig, and D. Johnson, Packet leashes: a defense against wormhole attacks in wireless networks, in *Proc. of IEEE INFOCOM* (2003).

21. D. Ganesan et al., Highly resilient, energy-efficient multipath routing in wireless sensor networks, *Mobile Comp. and Commun. Review*, Vol. 5 No. 4 (2002) pp. 10–24.

22. V. Gupta, S. Krishnaurthy, and M. Faloutsos, Denial of service attacks at the MAC layer in wireless ad hoc networks, in *Proc. of IEEE MILCOM* (2002).

23. N. Borisov, I. Goldberg, and D. Wagner, Intercepting mobile communications: the insecurity of 802.11, in *Proc. of the Seventh Annual International Conference on Mobile Computing and Networking*, Rome, (2001) pp. 180–189.

24. J.P. Hubaux, L. Butttan, and S. Capkun, The quest for security in mobile ad hoc networks, in *Proc. of ACM MOBIHOC* (2001).

25. D. Malan, H. Wlsh, and M. Smith, A public-key infrastructure for key distribution in TinyOS based on elliptic curve cryptography, in *Proc. of the First IEEE International Conference on Sensor and Ad Hoc Communications and Networks, SECON04* (2004).

26. G. Gaubatz, J.P. Kaps, and B. Sunar, Public key cryptography in sensor networks, in *Proc. of the First European Workshop on Security in Ad Hoc and Sensor Networks (ESAS 2004)*, 2004.

27. Q. Huang, J. Cukier, H. Kobayashi, B. Liu, and J. Zhang, Fast authenticated key establishment protocols for self-organizing sensor networks, in *Proc. of the 2nd ACM International Conference on Wireless Sensor Networks and Applications* (2003).

28. L. Butttyan and J.P. Hubaux, Rational exchange — a formal model based on game theory, in *Proc. of the 2nd International Workshop on Electronic Commerce*, (2001) pp. 114–126.

29. F. Ye, H. Luo, S. Lu, and L. Zhang, Statistical en-route filtering of injected false data in sensor networks, in *Proc. of IEEE INFOCOM*, (2004).

30. H. Yang, F. Ye, Y. Yuan, S. Lu, and W. Arbaugh, Toward resilient security in wireless sensor networks, in *Proc. of the International Symposium on Mobile Ad Hoc Networking and Computing, ACM* (2005).

31. J.F. Kurose and K.W. Ross, *Computer Networking — A Top-Down Approach Featuring the Internet*, Addison Wesley (2001).

32. R. Wattenhofer, L. Li, P. Bahl, and Y.M. Wang, Distributed topology control for power efficient operation in multihop wireless ad hoc networks, in *Proc. of IEEE INFOCOM*, (2001) pp. 1388–1397.

33. J. Aslam, Q. Li, and D. Rus, Three power-aware routing algorithms for sensor networks, *Wireless Communications and Mobile Computing*, Vol. 2 No. 3 (2003) pp. 187–208.

34. A. Woo and D.E. Culler, A transmission control scheme for media access in sensor networks, in *Proc. of ACM MOBICOM*, (2001) pp. 221–235.

35. E.S. Jung and N. Vaidya, A power saving mac protocol for wireless networks, in *Proc. of IEEE INFOCOM*, (2002) pp. 1756–1764.

36. S.R. Madden, M.J. Franklin, J.M. Hellerstein, and W. Hong, TAG: a tiny aggregation service for ad-hoc sensor networks, in *Proc. of OSDI*, (2002).

37. S.R. Madden, R. Szewczyk, M.J. Franklin, and D. Culler, Supporting aggregate queries over ad-hoc wireless sensor networks, in *Workshop on Mobile Computing and Systems Applications*, (2002) pp. 49–58.

38. Z. Yu and Y. Guan, A robust group-based key management scheme for wireless sensor networks, in *Proc. of the IEEE Wireless Communications and Networking Conference*, Vol. 4 (2005) pp. 13–17.

39. R. Min, M. Bhardwaj, N. Ickes, A. Wang, and A. Chandrakasan, The hardware and the network: total-system strategies for power aware wireless microsensors, in *Proc. of the IEEE CAS Workshop on Wireless Communications and Networking*, Pasadena, CA (2002).

40. K. Jones, A.Wadaa, S. Oladu, and L. Wilson, Towards a new paradigm for securing wireless sensor networks, in *Proc. of the 2003 Workshop on New Security Paradigms*, (2003) pp. 115–121.

41. L. Lazos and R. Poovendran, Cross-layer design for energy-efficient secure multicast communications in ad hoc networks, in *Proc. of IEEE International Conference on Communications*, Vol. 6, No. 20–24 (2004) pp. 3633–3639.

42. L. Lazos, J. Salido, and R. Poovendran, VP3: using vertex path and power proximity for energy efficient key distribution, in *Proc. of the IEEE Vehicular Technology Conference, VTC2004–Fall*, Vol. 2 (2004) pp. 1228–1232.

43. K. Jones, A.Wadaa, S. Oladu, L. Wilson, and M. Etoweissy, Towards a new paradigm for securing wireless sensor networks, in *Proc. of the New Security Paradigms Workshop*, (2003) pp. 115–121.

44. S. Slijepcevic and M. Potkonjak, On communication security in wireless ad hoc sensor networks, in *Proc. of the Eleventh IEEE International*

Workshop on Enabling Technologies: Infrastructure for Collaborative Enterprises (WETICE), Vol. 1 No. 1 (2002) pp. 139–144.

45. X. Qi and A. Ganz, Runtime security composition for sensor networks (SecureSense), in *Proc. of the IEEE Vehicular Technology Conference, VTC 2003–Fall*, Vol. 5 (2003) pp. 2976–2980.

46. Y. Zhang and W. Lee, Intrusion detection in wireless ad hoc networks, in *Proc. of ACM MOBICOM*, (2000) pp. 275–283.

47. L. Eschenauer and V.D. Gligor, A key-management scheme for sensor networks, in *Proc. of the 9th ACM Conference on Computer and Communication Security*, (2002) pp. 41–47.

48. W. Du, J. Deng, Y.S. Han, and P.K. Varshney, A pair-wise key predistribution scheme for wireless sensor networks, in *Proc. of ACM CCS*, (2003) pp. 42–51.

Index